MCQ and Theory on Soil & Water Conservation Engineering

(for M.Sc., PhD, ARS, JRF, SRF, JE etc.)

Other Related Books

Organic Agriculture	*J.C. Tarafdar*
Soil Survey Manual New Revised Edition	*USDA*
Soil and Plant Analysis	*C.S. Piper*
Salinity Tolerance in Plants: Methods, Mechanisms and Management	*B. K. Garg*
Alkali Wastelands Environment and Reclamation	*I.C. Gupta*
Soil and Plant Analysis Laboratory Manual, 2nd Edition	*J. Ryan*
Soil and Water Conservation in Semi-Arid Areas	*Norman W. Hudson*
Soil Salinity Assessment	*FAO*
Use of Saline Water in Agriculture	*I.C. Gupta*
Standard Methods for Analysis of Soil Plant and Water	*I.C. Gupta*
Salt Affected Soils : Reclamation and Management	*S.K. Gupta*
Nanotechnology	*J.C. Tarafdar*
Management of Saline & Waste Water in Agriculture	*S.K. Gupta*
Genesis and Management of Sodic (Alkali) Soils	*S.K. Gupta*
Soil Chemical Analysis	*M.L. Jackson*
Crop Production In Salt Affected Soils	*I.C. Gupta*
Salt Affected Soils: Reclamation and Management	*S.K. Gupta*
Soilborne Plant Pathogens Management of Diseases with Macro and Microelements	*A.W. Engelhard*
Nanofertilizers: Challenges and Prospects	*J.C. Tarafdar*
Salinity Tolerance in Plants: Methods, Mechanisms and Management 2nd Ed	*B. K. Garg*
Handbook of Saline and Alkali Soils Diagnosis Reclamation and Management	*S.K. Gupta*
Handbook of Saline and Alkali Soils Diagnosis Reclamation and Management	*S.K. Gupta*
Crop Production in Waterlogged Saline Soils	*S.K. Gupta*
Disaster Perception and Management	*T.S. Chouhan*
Measuring Agricultural Sustainability	*Kaushalya Ramachandran*

MCQ and Theory on Soil & Water Conservation Engineering

(for M.Sc., PhD, ARS, JRF, SRF, JE etc.)

Er. Ramesh Verma

(Ph.D., Research Scholar)

SWCE, Deptt. of Farm Engineering,

Institute of Agricultural Sciences, BHU, Varanasi

Prof. Munish Kumar

[M.Sc. (Ag), Ph.D., FSAP]

Department of Soil Conservation and

Water Management

Chandra Shekhar Azad University of

Agriculture and Technology

Kanpur -208002 (U.P.) India

SCIENTIFIC
PUBLISHERS

Published by

SCIENTIFIC PUBLISHERS (INDIA)
5 A, New Pali Road, P.O. Box 91
JODHPUR 342 001 (India)
E-mail: info@scientificpub.com
Website: www.scientificpubonline.com

ISBN: 978-93-90495-15-3
e-ISBN: 978-93-90495-19-1

Printed in India.

PREFACE

This objective book is intended to be used by students of Soil and Water Conservation Engineering, Irrigation & Drainage Engineering/ Irrigation & Water Management Engineering, Land and Water Management Engineering and Agricultural Engineering/Agricultural students of different colleges of Indian Universities. The purpose of this book is to make students excel in different exams like Agricultural Research Service (ARS), National Eligibility Test (NET), and PhD entrance test of different State Agricultural Universities/Central Agricultural Universities etc. Chapters have been identified according to the ARS exam pattern. Some useful information regarding various Iso terms, definitions, agricultural research institute, list of agricultural universities in India have been included for ready reference for users/readers. The authors would feel obliged if the readers points out both grammatical and technical error if any, in the book for rectifying in the successive editions.

— Er. Ramesh Verma

— Prof. Munish Kumar

ACKNOWLEDGEMENT

We are grateful for number of friends and colleagues in encouraging us to write a MCQ book on the subject and publish it. We extend our gratefulness to the head of Institutes for providing congenial environment to accomplish the task. The graces of the God have always blessed us and given us patience and power to overcome the difficulties, which helped in accomplishment of this new endeavour. Finally, we would like to acknowledge with gratitude, the support and love of our family members to complete the task.

We are also gratefully acknowledging support of our publisher and editors for publishing this book very timely.

— Er. Ramesh Verma

— Prof. Munish Kumar

CONTENT

Multiple Choice Question

Theory

CHAPTER 1

SOIL EROSION

1. **Which of the following crops is the most effective in controlling soil erosion:**
(A) Maize (B) Cotton
(C) Green gram (D) Pigeon pea

2. **Sheet erosion is caused by:**
(A) Wind (B) Glaciers
(C) Heavy rains
(D) Fast running rivers

3. **Deforestation causes due to:**
(A) Pollution (B) No floods
(C) Soil erosion (D) None of these

4. **Main indicator of soil erosion is:**
(A) Ozone layer
(B) Deforestation
(C) Afforestation
(D) Goats and grazing animals

5. **Contour bunding is done to check:**
(A) Sheet erosion
(B) Rill erosion
(C) Gully erosion
(D) Ravine Formation

6. **Eroded soils are:**
(A) Rich in plants nutrients
(B) Devoid of plant nutrients
(C) Unaltered in plant nutrients
(D) All of these

7. **Soil erosion can be reduced by:**
(A) Making dams
(B) Planting proper plants
(C) Reducing over grazing
(D) All of these

8. **Soil erosion can be prevented by:**
(A) Heavy rains (B) Over grazing
(C) Afforestation (D) Deforestation

9. **If there is a little slope in the field, the soil erosion can be prevented by:**
(A) Contour farming
(B) Contour terracing
(C) Contour strip cropping
(D) None of these

10. **The disadvantage of ploughing in arid and semi. arid zones is:**
(A) Loss of soil moisture
(B) Soil organisms are exposed
(C) Soil particles became smaller
(D) All of these

11. **Gullies are removed by**
(A) Levelling
(B) Pan breaking
(C) Ridge terracing
(D) All of these

12. **An important agrostological technique to check soil erosion in the initial stages is:**
(A) Contour farming
(B) Land retirement
(C) Basin listing
(D) Ley farming

13. **Which is the main causative factor for desertification?**
(A) Tourism
(B) Overgrazing
(C) Irrigated agriculture
(D) Developmental activities

14. **Soil erosion can be minimised through:**
(A) Increasing fertility
(B) Making the land sloppy
(C) Growing plants to form a soil cover
(D) Allowing herbivorous animals to graze excessively

15. To prevent wind and water erosion, the crop is sometimes harvested in such a way that a basal stub of the plant is left behind. This phenomenon is called:
(A) Mulching (B) Ley farming
(C) Basin listing (D) Strip cropping

16. Soil erosion in plains consequent to rainfall is initiated by:
(A) Rill erosion (B) Gully erosion
(C) Sheet erosion (D) All of these

17. The term riparian erosion is applied to:
(A) Slip erosion (B) Gully erosion
(C) Wind erosion
(D) Stream bank erosion

18. Bunds, dams and drains are constructed to prevent:
(A) Gully erosion (B) Slip erosion
(C) Sheet erosion
(D) Stream bank erosion

19. Soil erosion is a three phase phenomena:
(A) Detachments (B) Transportation
(C) Deposition (D) All

20. Gully erosion is the:
(A) Last stage of rill formation
(B) Advanced stage of rill formation
(C) Pre stage
(D) None of these

21. The sequence of water erosion is:
(A) Splash, Sheet, Rill, Gully
(B) Sheet, Gully, Rill
(C) Rill, Splash, Sheet
(D) Gully erosion, Splash, Sheet, Rill

22. Soil erosion which occurs under normal conditions are called as:
(A) Surface creep
(B) Geological erosion
(C) Both (A) and (B)
(D) Accelerated erosion

23. Rill erosion is also called as:
(A) Micro channel irrigation
(B) Major channel irrigation
(C) None of the above
(D) Both (A) and (B)
(E) None of these

24. Erosion of very fine particles is seen on account of:
(A) Saltation
(B) Suspension
(C) Surface creep
(D) Shifting of dunes

25. Soil erosion is the function of:
(A) Erosivity
(B) Erodibility
(C) Both (A) and (B)
(D) Porosity

26. Odum has described soil erosion as:
(A) Soil pollution
(B) Loss of feeding zone
(C) Loss of living stratum
(D) Creeping death of soil

27. A sloppy land on account of continued rainfall initially under goes:
(A) Gully erosion
(B) Sheet erosion
(C) Rill erosion
(D) All of these

28. The chief cause of accelerated erosion is:
(A) Over felling and overgrazing
(B) Excessive rains
(C) Wind storms
(D) All of the above

29. Wave erosion is more prevalent and well marked:
(A) Along rivers
(B) In water falls
(C) At the sea shore
(D) In all these places

30. Erosivity is the function of:
(A) Rainfall intensity
(B) Rainfall
(C) Runoff
(D) Temperature

31. Wind erosion will be higher from:
(A) A Barren sandy soil
(B) Clay soil
(C) Sandy soil
(D) Loam soil

32. **Wind velocity is less at:**
(A) Ground surface
(B) Above the ground
(C) Below the ground
(D) All of the above

33. **Wind velocity higher at:**
(A) 50 m height from the ground
(B) 80 m height from the ground
(C) 100 m height from the ground
(D) 150 m height from the ground

34. **Wind turbulence is increases with increase in:**
(A) Friction velocity (B) Velocity
(C) Pressure (D) Temperature

35. **Wind turbulence is greater on:**
(A) Rough surface
(B) Smooth surface
(C) Undulating surface
(D) None of the above

36. **The magnitude of wind turbulence is greater at:**
(A) Ground surface
(B) Rough surface
(C) Smooth surface
(D) None of the above

Answers Key

Q. Ans	Q. Ans	Q. Ans	Q. Ans	Q. Ans	Q. Ans	Q. Ans	Q. Ans	Q. Ans
1. D	2. C	3. C	4. D	5. A	6. B	7. D	8. C	9. C
10. A	11. A	12. D	13. B	14. C	15. A	16. C	17. D	18. A
19. D	20. A	21. A	22. B	23. A	24. A	25. C	26. A	27. C
28. A	29. C	30. A	31. A	32. A	33. A	34. A	35. A	36. A

CHAPTER 2

SOIL CONSERVATION

1. If the top soil has been eroded, we retire a land and grow over it:
   (A) Grasses
   (B) Lichens
   (C) Legumes
   (D) Blue green algae

2. Terracing is an effective method of soil conservation:
   (A) Plains (B) Hilly areas
   (C) Desert areas (D) Riverine areas

3. Soil conservation means:
   (A) Prevention of spread of desert
   (B) To check soil erosion by afforestation
   (C) To check soil erosion by wind and rains
   (D) All of the above

4. Soil conservation can best achieved by having:
   (A) Wind Screen
   (B) Low rainfall
   (C) Good plant covers
   (D) Restricted human activity

5. The soil binding is done in the best way by:
   (A) Algae (B) Grasses
   (C) Mycorhize (D) All of the above

6. When trees and shrubs are planted in long rows along streams, described as:
   (A) Wind breaks (B) Shelter belts
   (C) Soil binders (D) Basin blisters

7. Digging pits on the slopes just for the accumulation of runoff is called as:
   (A) Contour trenching
   (B) Contour terracing
   (C) Beach terracing
   (D) None of these

8. Which of the following prevents afforestation in deserts?
   (A) Low rainfall
   (B) Local resident
   (C) Unfertillity of soil
   (D) Goats and grazing animals

9. Rotation of crop is essential for:
   (A) Increasing fertility of soil
   (B) Epiphytes
   (C) Nematode disease
   (D) Agricultural growth

10. The fertility level of any soil is generally determined by:
    (A) Aeration and hydration
    (B) Humus and mineral contents
    (C) Humus, mineral and hydration level
    (D) All of these

11. The most common method used for cultivation on hills:
    (A) Bench terracing
    (B) Ridge terracing
    (C) Channel terracing
    (D) Levelling

12. The most significant and initial requirement for dry farming is:
    (A) Crop selection
    (B) Animal husbandry
    (C) Shallow ploughing
    (D) Development of textures

13. The soil permeability can be decreased due to blockers called:
    (A) Clods
    (B) Pans
    (C) Peds
    (D) None of these

14. The removal of weeds by ploughing is called:
(A) Tilling
(B) Mulching
(C) Fallowing
(D) Contour farming

15. Which of the following plants is used as windbreak?
(A) Delonix (B) Prosopis
(C) Mangifera (D) Eucalyptus

16. Side effect of heavy and constant grazing is the:
(A) Upper layers of soil are loosened and are eroded
(B) Lower compact layers make root penetration difficult
(C) Lower layers are made more compact and root respiration is interfered with
(D) All of the above.

17. Constant grazing and browsing in an area may ultimately result in the formation of:
(A) Desert
(B) Grassland
(C) Dense forest
(D) Bushy vegetation

18. The possible beneficial effect of the grazing animals is:
(A) Removal of wild plants
(B) Eradication of weeds
(C) Removal of wild animals
(D) Addition of their excreta to the soil

19. When young plants are grazed too heavily they get killed because:
(A) The roots are starved
(B) Of the mechanical injury to roots
(C) Most of the aerial parts are eaten away
(D) All of the above

20. The grazing animals can change the type of vegetation by:
(A) Cross pollination
(B) Selective grazing
(C) Bringing in plant pathogens
(D) Bringing the seeds of other plants

21. Soil conservation is the process where:
(A) Soil is aerated
(B) Soil erosion is allowed
(C) Soil is protected against loss
(D) Sterile soil is converted into fertile soil

22. The plants commonly sown for crop rotation are:
(A) Trigonella and Trifolium
(B) Cajanus and Dalbergia
(C) Dalbergia and trigonella
(D) Cajanus and Aeschynomene

23. In dry farming initially the fertility of the land can be increased by:
(A) Agrostology
(B) Basin listing
(C) Animal husbandry
(D) Growing grazing fields

24. The cutting of trees from the forests of hill near a catchment area:
(A) Will have no effect on causing floods in plain
(B) May cause flood in plains in rainy season
(C) Will benefit the mankind for more area for cultivation
(D) Will have no effect on climatic condition of that area

25. Grasses possess great regenerative power because they are:
(A) Fast growing
(B) Rhizomatous
(C) Dominate species
(D) Constantly grazed

26. If the same crop is repeatedly grown in a field:
(A) Water level in soil will recede`
(B) Nitrogen starvation many result
(C) Soil will become prone to diseases
(D) Specific mineral deficiency may arise

27. Gully development is accomplished under:
(A) Four (B) Three
(C) Two (D) One

28. Gabion structures are:
    (A) Flexible
    (B) Permeable
    (C) Both (A) and (B)
    (D) None of these

29. Land capability unit refers to the:
    (A) Classifying the land
    (B) Grade the land
    (C) Both
    (D) None of these

30. Class-III lands of LUCC are:
    (A) Moderately good for cultivation
    (B) Good for cultivation
    (C) Not good for cultivation
    (D) None of the above

31. Mulch tillage increases:
    (A) Moisture content in the soil
    (B) Presence of air
    (C) Flow of water
    (D) All of the above

32. Bunds are constructed for the purpose of:
    (A) Retaining water
    (B) Controlling soil loss
    (C) Both (A) and (B)
    (D) None of these

33. The drop spill way are Constructed at the depth height of:
    (A) More than 3 feet
    (B) More than 8 feet
    (C) More than 10 feet
    (D) More than 12 feet

34. Mulching helps in:
    (A) Weed control
    (B) Moisture conservation
    (C) Increasing soil fertility
    (D) Improvement of soil structure

Answers Key

Q. Ans	Q. Ans	Q. Ans	Q. Ans	Q. Ans	Q. Ans	Q. Ans	Q. Ans	Q. Ans
1. A	2. B	3. D	4. C	5. B	6. B	7. A	8. A	9. A
10. D	11. A	12. A	13. B	14. A	15. B	16. D	17. A	18. D
19. A	20. B	21. C	22. A	23. C	24. B	25. D	26. D	27. A
28. C	29. A	30. A	31. A	32. C	33. A	34. B		

CHAPTER 3

HYDROLOGY AND OPEN CHANNEL FLOW

1. Measurement and analysis of water is a science called as:
(A) Hydrometry (B) Hydrology
(C) Limnology (D) Hydrometeorology

2. Any channel which is supplied with water from canal and is maintained by farmers not by government is:
(A) Watercourse (B) Chak
(C) Module (D) Distributary

3. The structure in which discharge is constant and fixed within limits, irrespective of the fluctuations of the water levels of either the distributaries or of the water course or both called:
(A) Rigid modules
(B) Flexible modules
(C) Non. modular modules
(D) None

4. Warabandi or Osrabandi system distributes the water to farmers with main objective and requirement of:
(A) Demand (B) Supply
(C) Continuous (D) Equity

5. The kinematic viscosity of water at 200C is almost:
(A) 10.6 m^2/s (B) 10.4 m^2/s
(C) 10.2 m^2/s (D) 10.5 m^2/s

6. Darcy's law is valid in a porous flow media if Reynolds number is:
(A) >1 (B) <1
(C) >2000 (D) <2000

7. A sheet of water, which overflows a weir, is called as:
(A) Head (B) Nape
(C) Crest (D) Notch

8. Venturimeter works on the principle of:
(A) Darcy (B) Bernoulli
(C) Hooghoudt (D) Chezy

9. The size of the Parshall flume is specified by:
(A) Head in diverging section
(B) Head in converging section
(C) Throat width
(D) None of the above

10. The dimension of intrinsic permeability is:
(A) L (B) T
(C) LT.1 (D) L2

11. At critical flow condition, the specific energy is
(A) Maximum (B) Minimum
(C) Alter (D) No charge

12. The ratio of inertia force to gravitational force is called as:
(A) Froude number
(B) Weber number
(C) Reynolds number
(D) Pressure coefficient

13. For the same head, the discharge in contracted weir with two end contractions is:
(A) Less than supported weir
(B) More than supported weir
(C) Equals to supported weir
(D) None of the above

14. Weir under no ventilation condition comes under:
(A) Suppressed (B) Submerged
(C) Contracted (D) Free

15. **Most efficient hydraulic channel cross. section is:**
(A) Trapezoidal (B) Circular
(C) Rectangular (D) Semi. circular

16. **For a shallow wide rectangular channel, the ratio of wetted perimeter to hydraulic radius is:**
(A) Less (B) More
(C) Equal
(D) None of the above

17. **The length of hydraulic jump is calculated with sequential depths (y1 and y2) as:**
(A) Capacity (B) L= 6.9 (Y1.Y2)
(C) Spacing
(D) None of the above

18. **Hydraulic jump is an example of the flow type is:**
(A) Steady and non. uniform
(B) Unsteady and non. uniform
(C) Steady
(D) Unsteady

19. **The energy line is equal to gradient lines in case of flow through:**
(A) Pipelines
(B) Open channel
(C) Both (A) and (B)
(D) None of the above

20. **Bernoulli's principle is based on the conservation of:**
(A) Mass (B) Momentum
(C) Energy (D) Both (A) and (C)

21. **The Co. efficient of discharge of orifice is:**
(A) 0.62 (B) 0.99
(C) 0.95 (D) Both (B) and (C)

22. **The current meter is used to measure:**
(A) Velocity (B) Head
(C) Discharge (D) Pressure

23. **How much water column (cm) is in 1 bar pressure:**
(A) 1019.72 (B) 1034
(C) Both (A) and (B) (D) None of these

24. **What is the unit of bulk density:**
(A) g/cc
(B) Kg/cc
(C) Both (A) and (B)
(D) None of these

25. **A flow meter that is independent of fluid density:**
(A) Rotameter
(B) Orifice meter
(C) Electromagnetic flow meter
(D) None of these

26. **At critical flow condition in the channel the value of Froude number is:**
(A) Greater than 1 (B) Less than 1
(C) Equal to 1 (D) None of these

27. **A Parshall flume is based on the concept of:**
(A) Hydraulic jump (B) Hydraulic drop
(C) Uniform flow (D) Turbulent flow

28. **The cipoletti weir is contracted trapezoidal weir having side slope:**
(A) 2:1 (B) 1:2
(C) 4:1 (D) 1:4

29. **Reynolds number is defined as the ratio of:**
(A) Inertia to gravity force
(B) Inertia to elastic force
(C) Inertia to viscous force
(D) Viscous to inertia force

30. **Gauge pressure = absolute pressure ....**
(A) Vacuum pressure
(B) Atmospheric pressure
(C) Hydrostatic pressure
(D) None of the above

31. **Stokes law is used to find out:**
(A) Terminal velocity
(B) Drag coefficient
(C) Surface tension
(D) Viscosity

32. **An imaginary line connecting points of equal atmospheric pressure is called as:**
(A) Isobar (B) Isobath
(C) Isochar (D) Isopleth

33. Fluid flow in circular pipe is said to be Laminar if Reynolds number is:

(A) <1 (B) =1

(C) <2000 (D) >4000

34. Find the hydraulic mean depth(m) of lined canal water course rectangular in cross section with a bottom width 50 cm and depth of flow 25 cm:

(A) 0.125 (B) 0.126

(C) 0.127 (D) 0.128

35. If the Froude number of flow in an open channel is more than 1, then the flow is said to be:

(A) Critical

(B) Sub. critical

(C) Super-critical

(D) None of the above

36. For a fully developed laminar flow through a smooth pipe, the relationship between friction factor (f) and Reynolds number (Re) is:

(A) f α (Re) (B) f α (Re)-1

(C) f α (Re)2 (D) f α (Re)-2

37. If the density of a fluid changes from point to point in a flow regime, the flow is called:

(A) Steady flow

(B) unsteady flow

(C) Non. uniform flow

(D) Compressible flow

38. The viscosity of a Newtonian fluid depends primarily on X and to a lesser degree on Y. X and Y are:

(A) X= temperature, Y= flow velocity

(B) X= flow velocity, Y= pressure

(C) x= temperature, Y= pressure

(D) X= roughness of the surface across which the fluid flows, Y= flow velocity

39. While deriving the Chezy formula for uniform flow, it is assumed that there is a balance between:

(A) Gravity and Inertial forces

(B) Inertial and Viscous forces

(C) Frictional and Gravity forces

(D) Frictional and Inertial forces

40. The Co. efficient of discharge of Cipoletti weir is:

(A) 0.629 (B) 0.99

(C) 0.95 (D) Both (B) and (C)

41. A fluid in which shear stress is more than the yield value and proportional to the rate of shear strain is called:

(A) Newtonian fluid

(B) Ideal plastic fluid

(C) Non. Newtonian fluid

(D) Real fluid

42. The discharge (m^3/s) through a 900 V. notch for a head of 0.5 m and coefficient of discharge of 0.6 is:

(A) 0.25 (B) 0.50

(C) 0.65 (D) 0.75

43. The velocity head in the case of fluid flow is the:

(A) Kinetic energy per unit area

(B) Kinetic energy per unit flow area

(C) Kinetic energy per unit weight

(D) Kinetic energy per unit time

44. If the diameter of a pipe is halved flow of water in it experiences the increase in the head loss due to friction:

(A) Two Times

(B) Four Times

(C) Ten Times

(D) Sixteen Times

45. A dimensionless number representing the ratio of inertial forces to viscous forces in fluid flow is better known as:

(A) Stantun number

(B) Prandtl number

(C) Froude number

(D) Reynolds' number

46. The head over a V notch at the end of channel is 75 cm, if an error of 0.15 cm is possible in the measurement of head, then the percentage error in computing the discharge will be:

(A) 0.3 (B) 0.5

(C) 1.0 (D) 1.5

47. If discharge in two circular orifices O1 and O2 placed at depth of 20 cm and 80 cm below a constant level of water in a tank are equal, the ratio of the diameters of the orifices O1 and O2 is:
(A) 1:2 (B) 2:1
(C) 1.414:1 (D) 1:1.414

48. The time taken for a tank (filled to height 'h' above its flat base) to empty through an orifice in the base varies as the following power of 'h':
(A) 1 (B) 1/2
(C) -1/2 (D) -1

49. At sub – critical state of flow, the ratio of gravitational and inertial forces is:
(A) More than 1 (B) Less than 1
(C) Equal to 1 (D) None of these

50. Under open channel flow conditions, the hydraulic jump is formed when the Froude number is:
(A) Equal to one (B) More than one
(C) Less than one (D) None of the above

51. The equipment which continuously records the water level of the flowing stream is called as:
(A) Current Water
(B) Parshall flume
(C) Discharge recorder
(D) Stage level recorder

52. If the viscous forces are so strong, relative to the inertial forces, that viscosity plays a significant role in determining flow behaviour then flow in the channel will be:
(A) Turbulent
(B) Laminar
(C) Discontinuous flow
(D) None of these

53. Archimedes' principle is used in determining the:
(A) Shape of solid
(B) Surface area of irregular solids
(C) Centre of gravity of a floating body
(D) Volume of an irregular solid

54. If value of dy/dx =0 in the dynamic equation of gradually varied flow, then the flow is:
(A) Steady (B) Uniform
(C) Laminar (D) Non. Uniform

55. The velocity distribution across a pipe flowing full is:
(A) Uniform (B) Triangular
(C) Parabolic (D) Semi. circular

56. A rectangular channel 40 cm wide is carrying a discharge of 0.1 m_3/s at a depth of 20 cm State whether the flow will be:
(A) Critical (B) Subcritical
(C) Supercritical (D) None of these

57. If the saturated hydraulic conductivity of a soil is 1 m/day, the rate of water transmission across a rectangular area 100m long and 1 m height under hydraulic gradient will be:
(A) 1 m^3/day (B) 10 m^3/day
(C) 100 m^3/day (D) 1000 m^3/day

58. Local atmospheric pressure is measured by:
(A) Hydrometer (B) Hygrometer
(C) Barometer (D) Altimeter

59. Chezy formula assumes that:
(A) The fore resisting the flow per unit area of the stream bed is proportional to the velocity
(B) The effective component of the gravity forces causing the flow is equal to the total force of resistance
(C) The force resisting the flow per unit area of the stream bed is inversely proportional to the velocity
(D) None of the above

60. At alternate depth 1 m and 1 m, the value of Froude number is:
(A) Greater than 1
(B) Less than 1
(C) Equal to 1
(D) None of these

61. In measuring the rate of flow in irrigation canals using current meters, the measurements is taken at two locations below the water surface and the average of the two values provides an estimate of the average rate of flow. To obtain fairly accurate results the measurement points are located at:
(A) 0.1 and 0.9 of the depth of flow
(B) 0.2 and 0.8 of the depth of flow
(C) 0.3 and 0.7 of the depth of flow
(D) 0.4 and 0.6 of the depth of flow

62. The depth of flow over a sharp crested rectangular weir should not be more than about:
(A) Half the crest width
(B) Two thirds of the crest width
(C) Three fourth of the crest width
(D) The width of the weir

63. The gauge used to measure the depth of flow over a weir is located upstream of the weir at a distance of about:
(A) Twice the approximate head
(B) Thrice the approximate head
(C) Four times the approximate head
(D) Five times the approximate head

64. The appropriate device for the measurement of flow in irrigation furrows is:
(A) V. notch weir
(B) Dethridge meter
(C) Portable Parshall flume
(D) Submerged orifice

65. The most suitable device for measuring large flows in earth channels in a canal water distribution system is:
(A) In built Parshall flume
(B) Suppressed rectangular weir
(C) Broad crested rectangular weir
(D) Truncated flume.

66. The principle of working of a meter gate is:
(A) Free flow rectangular orifice
(B) Free flow circular orifice
(C) Submerged flow circular orifice
(D) Submerged flow rectangular orifice.

67. The permissible value of mean velocity of flow in an earth channel in loamy soil is:
(A) 40 cm/sec
(B) 60 cm/sec
(C) 80 cm/sec
(D) 1 m/sec

68. The most commonly used materials for lining canals in India and other South Asian countries are:
(A) Cement concrete
(B) Cement mortar
(C) Low density polyethylene film
(D) Bricks or tiles in sandwiched cement mortar

69. The minimum side slope of an earth channel for polyethylene lining is;
(A) 15:1 (B) 2:1
(C) 2.5:1 (D) 3:1

70. The policy for lining of watercourses for on.farm water distribution systems in irrigation command areas is to provide sub. outlets to serve every:
(A) 6 ha (B) 8 ha
(C) 10h. (D) 12 ha

71. In a programme on the improvement of an outlet command area the important activities include:
(A) Location of canal outlet at the most appropriate point
(B) Alignment of the water course along the ridge line or as close to it as possible
(C) Taking branch water courses to reach isolated field plots
(D) Rectangulation of land holdings, subject to topographic limitations and consent of beneficiary farmers.

72. To prevent damage of pipeline from the load acting on it, concrete pipes should be laid with their upper surface at a depth of about:
(A) 30 cm
(B) 45 cm
(C) 60 cm
(D) 75cm

73. The limiting operating pressure which can be sustained by a non. reinforced concrete pipe in irrigation water conveyance is:
(A) 4 m (B) 6 m
(C) 8 m (D) 10m

74. The most commonly used concrete pipe for underground pipeline water distribution system is:
(A) Pipes with bell ends
(B) Pipes with tongue and groove joints
(C) Pipes with faucet and spigot joints
(D) Pipes with collar joints.

75. The most important factor influencing the capacity of the delivery pipeline using loop system of pipe layout in a community tube. well irrigation system is:
(A) The diameter of the pipe
(B) The length of the pipeline
(C) Elevation of the weir in the water distribution chamber of the elevated water tank
(D) Elevation of the outlet farthest from the tube well.

Answers Key

Q. Ans	Q. Ans	Q. Ans	Q. Ans	Q. Ans	Q. Ans	Q. Ans	Q. Ans	Q. Ans
1. A	2. A	3. A	4. D	5. A	6. B	7. B	8. B	9. C
10. D	11. A	12. A	13. A	14. A	15. D	16. A	17. B	18. A
19. B	20. C	21. A	22. A	23. A	24. C	25. A	26. C	27. A
28. D	29. C	30. B	31. A	32. A	33. C	34. A	35. C	36. D
37. D	38. C	39. C	40. A	41. B	42. C	43. C	44. A	45. D
46. B	47. B	48. B	49. B	50. B	51. D	52. A	53. C	54. B
55. C	56. B	57. C	58. C	59. A	60. C	61. B	62. B	63. C
64. C	65. C	66. B	67. B	68. D	69. B	70. A	71. A	72. B
73. B	74. D	75. C						

CHAPTER 4

PUMPS AND WELLS

1. The geophysical method of ground-water exploration which is suitable for both case and uncased wells is:
(A) Electrical resistivity method
(B) Electric logging
(C) Gamma ray logging
(D) Seismic refraction surveying

2. Keeping the hydraulic properties of the aquifer as constant, doubling the diameter of the tube-well will increase the well yield by about:
(A) 7% (B) 9%
(C) 11% (D) 5%

3. In a homogeneous artesian aquifer, the total length of the well screen as a percentage of thickness of the aquifer being tapped should range between:
(A) 50-60% (B) 60-70%
(C) 70-80% (D) 80-90%

4. In the design of a tube well, the velocity of water entering a well screen should be kept about:
(A) 1 cm/sec (B) 3 cm/sec
(C) 5 cm/sec (C) 7 cm/sec

5. In drilling tube wells with direct rotary drill, the drilling speeds range from:
(A) 30 to 150 rpm (B) 150 to 300 rpm
(C) 300 to 500 rpm (D) 500 to 750 rpm

6. In reverse rotary drilling, the drilling fluid is recovered using:
(A) Reciprocating pump
(B) Centrifugal pump
(C) Rotary pump
(D) Propeller pump

7. In reverse rotary drilling, the desirable velocity of water inside the drill pipe to lift fine cuttings is about:
(A) 40 m/sec
(B) 60 m/sec
(C) 80 m/sec
(D) 100 m/sec

8. The drilling fluid used in air rotary drilling is:
(A) Clear water
(B) Water mixed with bentonite
(C) Water mixed with other additives
(D) Compressed air

9. The most suitable equipment for well drilling in hard rock areas is:
(A) Cable tool drilling
(B) Direct rotary drilling
(C) Down-the-hole air rotary drilling
(D) Reverse rotary drilling.

10. The optimum length of a well screen for a tube well in a confined aquifer should extend to:
(A) Full depth of the aquifer
(B) 50-60% of the depth of the aquifer
(C) 60-70% of the depth of the aquifer
(D) 70-80% of the depth of the aquifer

11. The slot width of a well screen should be limited to:
(A) 1.0-1.5mm (B) 1.5-5.0 mm
(C) 5.0-7.5 mm (D) 7.5-10.0mm

12. The open area of a well screen as a percentage of the surface area of the pipe is:
(A) 10% (B) 15%
(C) 20% (D) 25%

13. The drilling speed in a rotary drilling machine ranges from:
(A) 30-150 rpm (B) 150-250 rpm
(C) 250-500 rpm (D) 500-750 rpm

14. The ideal return velocity of the fluid in the angular space of a reverse rotary drilling machine is
(A) 20-40 m/min (B) 40-60 m/min
(C) 60-80 m/min (D) 80 -100 m/min

15. The type of drill bit most suitable to be used with a direct rotary drilling machine to cut through rock formations, with a tendency to become crooked is:
(A) Three-way drag bit
(B) Six-way drag bit
(C) Three-cone roller bit
(D) Gross roller bit

16. For fishing, a drill pipe dropped accidentally during rotary drilling, a suitable fishing tool is:
(A) Pipe locking tool
(B) Tapered tap
(C) Die overshot
(D) Circulating slip overshot

17. The maximum hook load capacity of a medium duty rotary drill ranges from:
(A) 8,000 to 13,000 kg
(B) 13,000 to 18,000 kg
(C) 18,000 to 23,000 kg
(D) 23,000 to 28,000 kg

18. The energy output of a windmill varies:
(A) Directly with the wind speed
(B) As the square of the wind speed
(C) As the cube of the wind speed
(D) Inversely proportional to the wind speed

19. Overload releases in electric motor starters are set to trip off when the motor current exceeds:
(A) The current rating of the motor
(B) 5% above the motor current
(C) 10% of the motor current
(D) 15% of the motor current

20. For optimum performance, the per cent submergence of an air lift pump is:
(A) 50% (B) 60%
(C) 70% (D) 80%

21. The commonly used limit of suction lift of horizontal centrifugal pumps used in lift in irrigation schemes is:
(A) 4m (B) 4.5m
(C) 5m (D) 5.5m

22. The usually adopted value of retention period of inflow into a sump well is:
(A) 3 min (B) 5 min
(C) 7 min (D) 9min

23. The minimum submergence of the pump bowl assembly of a turbine / submersible / mixed flow pump in a jack well is:
(A) 1 m (B) 2m
(C) 3 m (D) 4m

24. Archemedian screw is suitable to lift water from open water bodies to heights rang in from:
(A) 0.5 to 1.2 m (B) 1.2 to 3 m
(C) 2 to 3 m (D) 3 to 4 m

25. The gas consumption of a biogas engine per brake horse power per hour is:
(A) 0.25 m^3 (B) 0.45 m^3
(C) 0.65 m^3 (D) 0.85 m^3

26. The minimum wind speed required for the operation of a windmill usually ranges from:
(A) 4-6 km/hr (B) 6-8 km/hr
(C) 8-10 km/hr (D) 10-12 km/hr

27. The minimum clearance between the bottom of a vertically driven turbine type pump and the floor of the jack well is:
(A) 1 m (B) 1.5 m
(C) 3 m (D) 2.5 m

28. The minimum spacing between each pumping unit in the power range of 30 to 60 h.p. is:
(A) 60 cm (B) 70 cm
(C) 80 cm (D) 90 cm

29. The common type of pipe used for rising main in lift irrigation scheme is:
    (A) PVC (B) Cast-iron
    (C) RCC (D) Mild steel

30. Pipes specials commonly used in lift irrigation schemes are generally made of:
    (A) Cast iron (B) Mild steel
    (C) PVC (D) RCC

31. The total static head per stage of a lift irrigation system is limited to about:
    (A) 30 m (B) 40m
    (C) 50 m (D) 60m

Answers Key

Q. Ans	Q. Ans	Q. Ans	Q. Ans	Q. Ans	Q. Ans	Q. Ans	Q. Ans	Q. Ans
1. C	2. C	3. C	4. B	5. A	6. B	7. B	8. D	9. C
10. D	11. B	12. B	13. A	14. B	15. D	16. D	17. B	18. C
19. C	20. B	21. B	22. B	23. B	24. A	25. B	26. B	27. B
28. B	29. C	30. A	31. C					

CHAPTER 5

SOIL-PLANT-WATER RELATIONSHIPS

1. If bulk and particle densities of soil are 1.35 and 2.7 respectively. The void ratio is:

(A) 1 (B) 0.25
(C) 0.50 (D) 0.44

2. If saturation soil moisture content is 35% and bulk density of soil is 1.2 g/cc. Then the porosity is:

(A) 0.35 (B) 0.25
(C) 0.65 (D) 0.20

3. Discharge through porous material is directly proportional to the .... Power of radius of porous materials:

(A) 2nd (B) 6th
(C) 5th (D) 4th

4. The Bulk density is............. . Than the particle density:

(A) Less (B) More
(C) Equal (D) All of the above

5. Casgrande apparatus is used in measuring:

(A) Plastic limit (B) Liquid limit
(C) Solid limit (D) All of the above

6. A soil sample with porosity of 40%, its void ratio is:

(A) 0.06 (B) 0.28
(C) 0.40 (D) 0.66

7. Void ratio is defined as the ratio of:

(A) Volume of void to total volume
(B) Volume of void to volume of solid
(C) Volume of void to volume of water
(D) None of these

8. In the design of an agitator vessel with model volume V_1 and prototype volume V_2, the scale –up ratio is given by:

(A) V_1/V_2 (B) V_2/V_1
(C) $(V_2/V1)^{1/2}$ (D) $(V_2/V_1)^{1/3}$

9. A soil has a void ratio 0.75 and a specific gravity 2.66. The value of critical hydraulic gradient at which quick sand condition will occur is:

(A) 0.95 (B) 1.05
(C) 2.09 (D) 6.64

10. A completely saturated clay soil has particle density 2600 kg m^{-3} and bulk density 1400 kg m^{-3}. It has a moisture fraction of 40% on volume basis The porosity of the soil is:

(A) 0.46 (B) 0.54
(C) 0.74 (D) 0.86

11. If bulk density of a particulate material is 600 Kg m^{-3} and true density is 1000 kg m^{-3}. The porosity of the material is:

(A) 20% (B) 40%
(C) 60% (D) 80%

12. A saturated soil sample has 42.2% water content and unit weight 2.69. The void ratio of the soil sample will be:

(A) 0.784 (B) 0.478
(C) 0.874 (D) 0.087

13. The most desirable soil structure is:

(A) Granular structure
(B) Single grain structure
(C) Massive structure
(D) Compound structure

14. Among the predominant soil groups of India the largest area is under:

(A) Laterite soils
(B) Alluvial soils
(C) Red soils
(D) Black soils

15. **In India the maximum land area under Alfisols is in:**
(A) Karnataka
(B) Tamilnadu
(C) Andhra Pradesh
(D) Maharashtra

16. **In general, the major component of water requirement of crops is:**
(A) Pre-sowing irrigation
(B) Evapotranspiration
(C) Leaching requirement
(D) Deep percolation from crop root zone

17. **Generally, the most significant factor in determining the productivity of a crop is:**
(A) Non-availability of adequate soil moisture in the root zone at the critical stages of their growth
(B) Crop variety
(C) Nutrient availability
(D) Pests and diseases

18. **The most important factors influencing evaporation from land surfaces is:**
(A) Degree of saturation of soil surface
(B) Temperature of air and soil
(C) Humidity and wind velocity
(D) Density of vegetative cover

19. **Amongst the four distinct growth stages of seasonal crops the maximum value of crop coefficient Kc is during:**
(A) Initial stage
(B) Crop development stage
(C) Mid-season stage
(D) Late season stage

20. **The most important factor influencing the evapotranspiration of a crop is:**
(A) Physiology of the crop
(B) Stage of growth of the crop
(C) Degree of crop cover
(D) The location from where the data are obtained

21. **The principal meteorological factor influencing evapotranspiration is:**
(A) Air temperature (B) Solar radiation
(C) Humidity (D) Wind speed

22. **The more commonly used empirical formulae for estimating evapotranspiration, namely Blaney-Criddle method, Thornthwaite method and the Hargreaves method have been calibrated for:**
(A) Daily calculations
(B) Hourly calculations
(C) 10-day and monthly calculations
(D) Seasonal calculations

23. **Amongst the most commonly used empirical formulae to estimate evapotranspiration, the one which has been found to provide reasonable approximation at most locations is:**
(A) Blaney-Criddle formula
(B) Thornthwaite formula
(C) FAO Radiation method
(D) Hargreaves' alternative method

24. **A soil solution of pH 7 is called a:**
(A) Saline soil
(C) Alkali soil
(B) Acid soil
(D) Neutral soil

25. **A dispersed clay soil has:**
(A) High infiltration
(B) Low infiltration
(C) Good workability
(D) Good internal drainage

26. **Cations are attached to the surface of the clay particles by**
(A) Absorption
(B) Adsorption
(C) Adhesion
(D) Cohesion

27. **Saline soil can be reclaimed by:**
(A) Leaching
(B) Scrapping
(C) Adding gypsum
(D) By growing salt-tolerant crops

28. **In non-saline alkali soil, the exchangeable sodium percentage is:**
(A) <5 (B) 5-10
(C) 10-15 (D) >5

29. Clay soils are chemically active because they:
   (A) Have high specific surface
   (B) Are colloidal
   (C) Adsorb cations
   (D) Have porosity

30. An important natural factor causing soil salinity is:
   (A) High temperature
   (B) High evaporation
   (C) Low precipitation
   (D) Weathering of parent rock

31. Among the Indian Rivers, the water salinity is lowest and safest in:
   (A) Ganga
   (B) Yamuna
   (C) Kosi
   (D) Tungabhadra

32. In a furrow-irrigated saline land the crops should be sown:
   (A) On the ridge
   (B) Facing east
   (C) Facing north
   (D) At the furrow bottom

Answers Key

Q. Ans	Q. Ans	Q. Ans	Q. Ans	Q. Ans	Q. Ans	Q. Ans	Q. Ans	Q. Ans
1. A	2. A	3. D	4. A	5. C	6. D	7. D	8. D	9. A
10. A	11. B	12. A	13. A	14. C	15. C	16. B	17. A	18. D
19. C	20. B	21. B	22. C	23. D	24. D	25. B	26. B	27. A
28. D	29. A	30. B	31. C	32. D				

CHAPTER 6

IRRIGATION ENGINEERING

1. Over-all mineral level of soil recedes due to:
(A) Leaching only
(B) Utilization by plants
(C) Both (A) and (B)
(D) Chemical reactions

2. Upper Bari Doab canal is constructed on the river:
(A) Beas (B) Indus
(C) Ravi (D) Chenab

3. The empirical method for computing the consumptive use of a crop using the mean monthly temperature and day light hours is:
(A) Thornthwaite (B) Hargreaves
(C) Blaney Criddle (D) Lowry Johnson

4. Tolerable limit of water spreading for cotton crop to dispose water from field is:
(A) 7 days (B) 1 day
(C) 2 days (D) 3 days

5. .... Refers to the difference between field capacity and permanent wilting point:
(A) Available water
(B) Hygroscopic water
(C) Wilting point
(D) None of the above

6. .... Equation is the alternative of the FAO Penman Monteith equation:
(A) Thorntwaite (B) Hargreaves
(C) Blaney Criddle (D) Both (A) and (C)

7. Penman Monteith equation is the latest formula for estimation of:
(A) Reference ET (B) PET
(C) Evaporation (D) None of these

8. The cut. back stream concept cannot be used in:
(A) Graded furrows laid along the slope
(B) Contour furrows laid on gently sloping grade
(C) Level furrows on heavy soils
(D) Corrugations

9. The minimum land slope required in land irrigated with surface methods of water application is:
(A) 0.01% (B) 0.05%
(C) 0.10% (D) 0.15%

10. The pressure head in a sprinkler system is converted into velocity head at:
(A) The sprinkler head
(B) The sprinkler jet
(C) The nozzle
(D) The riser pipe

11. The size of the furrow stream usually varies from:
(A) 0.1 to 0.5 litres/sec
(B) 0.5 to 1.0 litres/sec
(C) 0.5 to 2 litres/sec
(D) 0.5 to 2.5 litres/sec

12. In sub-irrigation water reaches the plant roots through:
(A) Deep percolation
(B) Lateral spread of water from open ditches
(C) Capillary action
(D) Subsurface flow

13. The controllable variable influence in advance of the water front in an irrigation border is:

(A) Slope of the border strip
(B) Infiltration characteristics of the soil
(C) Size of the irrigation stream
(D) Hydraulic resistance offered by field surface roughness and vegetative are tardance

14. **The internal diameter of a micro tube ranges from:**
(A) 0.3 to 0.7 mm (B) 0.7 to 1.2 mm
(C) 1.7 to 2.2 mm (D) 2.2 to 2.7 mm

15. **The minimum time required for the settling of mineral particles, larger than 80 microns, in a settling basin is:**
(A) 5 min (B) 10 min
(C) 15 min (D) 20 min.

16. **The overland flow in an irrigation border strip is a case of:**
(A) Steady flow
(B) Unsteady flow
(C) Steady flow with decreasing discharge
(D) Unsteady flow with decreasing discharge

17. **Amongst the top 20 countries with large-scale adoption of sprinkler irrigation, the maximum percentage coverage under sprinklers is in:**
(A) USA (B) Saudi Arabia
(C) Libya (D) United Kingdom

18. **The major factor influencing the efficiency of a well designed sprinkler irrigation system is:**
(A) Operating pressure
(b)Wind velocity
(C) Sprinkler spacing
(D) Diameter of sprinkler lateral

19. **In a sprinkler irrigation system the most commonly used pipe for a permanently laid mainline is:**
(A) Asbestos cement pipe
(B) Aluminium pipe
(C) Steel pipe
(D) PVC pipe

20. **Globally, the maximum area under drip irrigation is in:**
(A) Israel
(B) United States of America
(C) Saudi Arabia
(D) France

21. **The diameter of the area covered by a rotating type permanent overhead type sprinkler system ranges from:**
(A) 15 to 20 m (B) 20 to 30 m
(C) 30 to 45 m (D) 40 to 50 m

22. **The minimum satisfactory value of uniformity coefficient under normal conditions of pressure and wind velocity is:**
(A) 80% (B) 85%
(C) 90% (D) 95%

23. **The maximum water application rate of a sprinkler in light sandy loam soils on land with slope ranging from 1 to 5% is:**
(A) 1.5 cm/hr (B) 2 cm/hr
(C) 2.5 cm/hr (D) 3 cm/hr

24. **The width of land covered by a perforated spray type sprinkler system ranges from:**
(A) 4 to 8m (B) 5 to 10
(C) 6 to 15 m (D) 8to20m

25. **In drip irrigation, water flows from the emission points through the soil by:**
(A) Infiltration
(B) Percolation
(C) Capillarity
(D) Capillarity and gravity

26. **The largest cropped area covered by drip irrigation in India is under:**
(A) Plantation crops
(B) Orchards
(C) Vegetable crops
(D) Fibber crops

27. **Ant damage in drip laterals can be controlled by using pipes not less than:**
(A) 6 mils. (B) 10 mils
(C) 15 mils. (D) 20 mils

28. **High density polyethylene pipe is produced at pressures of about:**
(A) 20 kg/sq. cm (B) 30 kg/sq. cm
(C) 40 kg/sq. cm (D) 50 kg/sq. cm

29. Under normal working condition, the maximum rate of emission by drippers is:
(A) 6 litres/hr (B) 10 litres/hr
(C) 15 litres/hr (D) 20 litres/hr

30. In a media filter, the size of the filter media (crushed granite, silica or quartz) is used to sieve the particle sizes is ranging from:
(A) 0.3 to 0.7 mm (B) 0.7 to 1.2 mm
(C) 1.2 to 1.7 mm (D) 1.7 to 2.2 mm

31. The pressure drop usually allowed in a media filter is about:
(A) 1m (B) 2m
(C) 3 m (D) 4m

32. The usual mesh size of the screen hole in a screen filter is:
(A) 80 mesh (B) 100 mesh
(C) 120 mesh (D) 140 mesh

33. In the chlorine treatment of irrigation water, the quantity of bleaching powder added in the water source is:
(A) 1mg/litre (B) 2 mg/litre
(C) 3 mg/litre (D) 4mg/litre

34. The minimum recommended earth cover for the mains and sub-mains in a permanently laid drip irrigation system for pipe sizes ranging from 6 to 10 cm diameter is:
(A) 40 cm (B) 50cm
(C) 60 cm (D) 75cm

35. The minimum specific gravity of the suspended particle which can be removed by a hydro-cyclone filter is:
(A) 1.0 (B) 1.2
(C) 1.4 (D) 1.6

Answers Key

Q. Ans	Q. Ans	Q. Ans	Q. Ans	Q. Ans	Q. Ans	Q. Ans	Q. Ans	Q. Ans
1. C	2. C	3. C	4. D	5. A	6. B	7. A	8. C	9. B
10. C	11. D	12. C	13. C	14. B	15. C	16. D	17. C	18. D
19. C	20. B	21. C	22. B	23. C	24. C	25. D	26. B	27. C
28. A	29. C	30. B	31. B	32. C	33. B	34. C	35. D	

CHAPTER 7

SOIL AND WATER CONSERVATION

(ENGINEERING ENTRANCE EXAM QUESTIONS)

1. Darcy's law is valid under the condition of:
   (A) Laminar flow with Reynold No. > 10
   (B) Reynold No. <1.0
   (C) Reynold No. > 1000
   (D) Study uniform flow

2. Ratio of volume of water added or removed directly from the saturated aquifer to the resulting change in volume of aquifer below the water table is called:
   (A) Apparent specific yield
   (B) Specific yield
   (C) Storage coefficient
   (D) Specific storage

3. A saturated soil sample has 42.2 per cent water content and unit weight 2.69. The void ratio of the soil sample will be:
   (A) 0.784 (B) 0.478
   (C) 0.874 (D) 0.087

4. Soil texture refers to :
   (A) Arrangement of soil particles
   (B) Size of soil particles
   (C) Colour of soil particles
   (D) None of the above

5. Cohesion is:
   (A) Attraction of similar molecules
   (B) Attraction between two different molecules
   (C) Both (A) & (B)
   (D) None of the above

6. Permeability of Clay soil as compared to sandy soil is:
   (A) Equal (B) More
   (C) Lesser (D) Difficult to say

7. If T is return period, then the term (T-1)/T give the probability of:
   (A) Occurrence of event.
   (B) Non-occurrence of event
   (C) Both (A) and (B)
   (D) None of the above

8. Metric potential is the result of phenomena of :
   (A) Adhesion
   (B) Capillarity
   (C) Both (A) & (B)
   (D) None of the above

9. Metric potential is always:
   (A) Positive
   (B) Negative
   (C) A sometime positive sometime negative
   (D) None

10. Osmotic potential is attributed to the presence of :
    (A) Solute in the soil.
    (B) Soil moisture tension in the soil
    (C) Both (A) & (B)
    (D) None of the above

11. At optimum soil moisture content, the dry density of soil will be :
    (A) Maximum
    (B) Minimum
    (C) Average
    (D) Zero

12. Tensiometer can effectively measure soil moisture tension in the range of :
    (A) 0.0 to 1 atm (B) <0.8 atm.
    (C) 0 to 15 atm (D) 15 to 33

13. Readily available soil moisture to plants in the soil profile (root zone) is approx. Equal to:
    (A) 100 per cent of available water holding capacity
    (B) 50 per cent of available water holding capacity
    (C) 10 per cent of available water holding capacity
    (D) 50 per cent of field capacity
14. The capillary water in the soil is mostly held between tension of about :
    (A) 0 to 0.01 bar
    (B) 1/3 bar to 15 bar
    (C) bar to 33 bar
    (D) 33 bar to 100 bar
15. Adhesion is :
    (A) Attraction of similar molecules.
    (B) Attraction between two different molecules
    (C) Both (A) & (B)
    (D) None of the above
16. A Centrifugal Pump running with 1450 RPM discharges 20 lps at 30 m total head. The specific speed of the pump will be :
    (A) 12 (B) 24
    (C) 16 (D) 20
17. Metric potential in soil profile can be measured using :
    (A) Hygrograph (B) Tensiometer
    (C) Potentiometer (D) Luxneter
18. Wheat supplied with similar quantity of water will cause greater uptake in:
    (A) Alkali soil (B) Sodic soils
    (C) Normal soils (D) Saline soils
19. An instrument used for measurement of saturated hydraulic conductivity of soils is:
    (A) Conductivity meter
    (B) Manometer
    (C) Permeameter
    (D) Hýdrometer
20. Rate of water loss from a short green grass which is never short of water is Known as:
    (A) Evapotranspiration
    (B) Consumptive use
    (C) Potential evapotranspiration
    (D) Transpiration
21. USWB Class A pan evaporimeter has diameter of about:
    (A) 100 cm (B) 120 cm
    (C) 150 cm (D) 175 cm
22. The ratio of volume of voids to the total soil volume is called:
    (A) Void ratio
    (B) Porosity
    (C) Dry bulk density
    (D) Wet bulk density
23. Keeping other factors constant, doubling diameter of tube well will increase the discharge by:
    (A) 11% (B) 20%
    (C) 50% (D) 100%
24. The wells in which the water level remains at the water table level are:
    (A) Non-artesian wells
    (B) Flow in artesian well
    (C) Non-flowing artesian well.
    (D) Confined well
25. The relation between duty and delta is:
    (A) Δ= 864 B/D (B) Δ=860 B/D
    (C) Δ=864D/B (D) None of the above
26. If the Impellor speed of centrifugal pump is doubled, the power consumption will be:
    (A) The same (B) 4 times
    (C) 8 times (D) 16 times
27. Modified Penman method for computing Potential evapotranspiration includes:
    (A) Energy terms.
    (B) Aerodynamic terms
    (C) Both (A) & (B)
    (D) None of the above
28. Water flow through a 1.2 m long cylindrical soil column having 650 cm^2 cross sectional area, is 800 litres per minute. If the hydraulic head is 1.5m, the

hydraulic conductivity of the soil will be:

(A) 16 cm/sec (B) 24 cm/sec
(C) 30 cm/sec (D) 36 cm/sec

29. Combined use of surface and ground water in an irrigation project is known as:
(A) Integrated use
(B) consumptive use
(C) Conjunctive use
(D) Bottom up use

30. Crown root initiation is one of the critical stages susceptible to moisture stress in:
(A) Chickpea (B) Pigeonpea
(C) Cowpea (D) Wheat

31. Time considered for estimating drainage coefficient is :
(A) 24 minutes
(B) 24hrs
(C) 48 hrs
(D) None of the above

32. In humid areas drainage is planned according to:
(A) Rainfall
(B) Humidity
(C) Number of rainy days
(D) None of the above

33. Field reconnaissance is done to prepare:
(A) Work outline for detailed investigations
(B) To develop general plan and estimates
(C) Both (A) and (B)
(D) Preparation of construction plan & specifications

34. Normally the root zone depth considered to be kept free from ground water level below:
(A) 0.5m (B) 1.5 m
(C) 3m (D) 4m

35. Which of the following crops is most susceptible to water logging?
(A) Vegetables (B) Field crops
(C) Paddy (D) Fruits

36. Usually the salinity problem occurs in:
(A) Sub-humid region
(B) Arid region
(C) Humid region
(D) tropical region

37. Infiltration is measured by:
(A) Cylindrical metal rings
(B) USWB class A pan
(C) Lysimeter
(D) Rain gauge

38. Evaporation from water surface is the process by which liquid on free surface is transformed into:
(A) Rainfall (B) Clouds
(C) A solid state (D) A gaseous state

39. Evaporation from a free water surface is measured by:
(A) Lysimeter
(B) Infiltrometer
(C) USWB class A Pan
(D) Anemometer

40. Overall project irrigation efficiency in Govt. owned irrigation projects in India is about:
(A) 5% (B) 30%
(C) 70% (D) 80%

41. In normal condition, water loss from surface irrigation varies from:
(A) 25 to 30% (B) 30 to 45%
(C) 10 to 15% (D) 15 to 20%

42. Canal lining is essential to check the:
(A) Seepage losses
(B) Growth of weeds
(C) Evaporation losses
(D) Flow rate of canal

43. PIM refers to:
(A) Programme of integrated management
(B) Participatory irrigation management
(C) Pressure irrigation management
(D) Private irrigation management

44. Water meter is used for measuring the:
(A) stream current (B) Pipe flow
(C) Runoff (D) Channel flow

45. **Which of the following is not related to irrigation?**
(A) Check gate
(B) Watercourse
(C) Turnout
(D) Coshocton wheel

46. **Which of the following terms is not related to drainage?**
(A) Venturi meter
(B) Mole
(C) 20-40 rule
(D) Water logging

47. **Drainable water is the:**
(A) Hygroscopic water
(B) Capillary water
(C) Gravitational water
(D) Perched water

48. **Salinity problem can be controlled by:**
(A) sub-surface drainage
(B) surface drainage
(C) interceptor drains.
(D) tillage operation

49. **In soil, the available form of water to the plant is:**
(A) gravitational water
(B) hygroscopic water
(C) Capillary water
(D) ground water

50. **Ground water refers to the water in:**
(A) Surface pond
(B) Aquifer
(C) Soils
(D) capillaries

51. **Confined aquifer is also known as:**
(A) Water table aquifer
(B) Artesian-aquifer
(C) Semi-confined aquifer
(D) Perched aquifer

52. **Unconfined aquifer is also known as:**
(A) Water table aquifer
(B) Artesian aquifer
(C) Semi-confined aquifer
(D) Perched aquifer

53. **Perched aquifer is located:**
(A) Below the main water table
(B) Above the main water. table
(C) Just below the water table.
(D) None of the above

54. **Drilling fluid is:**
(A) Water
(B) Water plus chemical
(C) Mud-Water mixed with bentonite clay
(D) Water mixed with cement

55. **Use of canal water and ground water in the same irrigation project command:**
(A) Integrated use (B) Conjunctive use
(C) Consumptive use (D) Double use

56. **Semi-confined aquifer is also known as:**
(A) Semi-unconfined aquifer
(B) Leaky aquifer
(C) Perched Aquifer
(D) Water table aquifer

57. **The formation which contains sufficient water but has very small quantity to transfer the water is called:**
(A) Aquifuse (B) Acquiclude
(C) Aquitard (D) Blind aquifer

58. **Porosity of soil formation is the ratio of:**
(A) Volume of formation to the volume of voids
(B) Volume of voids to the volume of formation
(C) Water volume to the soil volume
(D) Percent micro pores in the formation

59. **Piezometer are installed for measuring pressure head in:**
(A) Perched aquifer
(B) Confined aquifer
(C) Unconfined aquifer
(D) Open-well

60. **Well log is prepared during:**
(A) Gravel packing
(B) Well development
(C) Bore drilling
(D) Strainer installation

61. Well development refers to:
(A) Removal of well incrustation
(B) Increase in well discharge
(C) Removal of final particle from the area around the well screen
(D) Well testing

62. Pumping test in the well is continued:
(A) Till steady state
(B) 24 hours
(C) For 48 hours
(D) For 82 hours

63. Any statement by which output of a system can be determined, given the policy, the initial value of the state variable and system parameters is known as:
(A) State variable
(B) Objective function
(C) System parameter
(D) System constraint

64. The simplex procedure is used to solve general maximization problem in:
(A) Dynamic programming.
(B) Non-linear programming
(C) Linear programming
(D) Analog simulation

65. River flows are described as:
(A) Stochastic event.
(B) Deterministic event
(C) Constrained event
(D) Static event

66. For Water resource development project, the sum of all allocated benefits is divided by total cost to arrive at:
(A) Benefit-cost ratio
(B) Cost-benefit ratio
(C) Project viability ratio
(D) Project benefit Index

67. Bucket capacity of tipping bucket rain gauge is:
(A) 0.25 mm of rainfall
(B) 2.5 cm of rainfall
(C) 12.7 mm of rainfall
(D) 5.0 cm of rain

68. The arithmetic mean method for computing missing rainfall data is suitable when normal annual precipitations of various stations are within:
(A) 10% of the normal annual precipitation of missing station
(B) 20% of the mean annual precipitation at the missing station
(C) 10% of the mean annual precipitation at the missing station
(D) 15% of the normal rainfall at missing station

69. The formula for recurrence interval return period is given by:
(A) T=1/N
(B) $T = (P)^{1/2}$
(C) T= (N+1)/M
(D) none of the above
Where, T= Recurrence interval, N= Year, M= Rank

70. The imaginary line of the ground water table from which the water table slope downward away from both side is called as :
(A) Ground water divide
(B) Drainage divide
(C) Boundary line of hydrologic unit
(D) None of the above

71. The value of circulatory ratio of watershed varies from:
(A) 0.2 to 0.8
(B) 04 to 1.0
(C) 1.0 to 1.5
(D) 1.5 to 2.0

72. Runoff coefficient is dependent on:
(A) Nature of soil surface
(B) Land use
(C) Rainfall intensity
(D) All of the above

73. Electromagnetic method of stream measurement is based on the principle of:
(A) Faraday's
(B) Ohm's law
(C) Kirchhoff's law
(D) Both (A) & (C)

74. **The kirpich formula estimates the Tc is given by:**
(A) $Tc=0.02(L/H^{3/2})^{0.77}$
(B) $Tc=0.01947\,L^{0.77}\,S^{-0.385}$
(C) $Tc=0.02(LS)^{-0.385}$
(D) $Tc=0.02(LS)^{0.77}$

75. **By increasing land slope 4 times the transportation of particle is increased by:**
(A) 4 times (B) 8 times
(C) 16 times (D) 32 times

76. **At a greater land slope, the possibility of:**
(A) soil splash is more
(B) soil-detachment is less
(C) soil detachment is more
(D) Both (A) & (C)

77. **Slope length affects the erosion mainly by:**
(A) increasing flow velocity for shorter duration
(B) decreasing flow velocity for shorter duration
(C) increasing flow velocity for longer duration
(D) none of the above

78. **The limit of slope length at which soil erosion begins is called as:**
(A) optimum slope length
(B) Critical slope length
(C) allowable slope length
(D) none of the above

79. **Transportation of soil particles under splash erosion will be greater on:**
(A) level land surface
(B) level land without cover
(C) Irregular surface
(D) Uniform sloppy land surface

80. **Soil detachment by raindrop is independent of:**
(A) Land slope
(B) Soil colour
(C) Soil depth
(D) Soil texture

81. **Rill erosion is also known as:**
(A) Micro channel erosion
(B) Path erosion
(C) Micro erosion
(D) Gully erosion

82. **Sheet flow is generated, when:**
(A) Land slope is steep
(B) Land surface is rough
(C) Land surface is smooth with uniform slope
(D) Both (A) & (B)

83. **Rill erosion usually begins in the:**
(A) Lower part of land slope
(B) Middle of land slope
(C) Upper part of land slope
(D) Entire length of land slope

84. **Sediment yield will be higher from the watershed dominated by:**
(A) Splash or raindrop erosion
(B) Sheet erosion
(C) Rill erosion
(D) Both (A) & (B)

85. **Soil detachment in raindrop erosion takes place due to:**
(A) K.E. of raindrop (B) P.E. of raindrop
(C) Running flow (D) Land slope

86. **Detachment of soil particle by flowing water varies as:**
(A) Square root of its velocity
(B) Power 1.5 of its velocity
(C) Square of its velocity
(D) Power three of its velocity

87. **Transportation ability of flowing water varies as:**
(A) Fifth power of its velocity
(B) Square root of its velocity
(C) Square of its velocity
(D) Fourth power of its velocity

88. **Organic erosion occurs in the form of:**
(A) Phytogenic erosion
(B) Zoogenic erosion
(C) Attrition
(D) Both (A) & (B)

89. Erosion under shifting cultivation, deforestation & cultivation on steep slopes without protective measures is associated to:
(A) Anthropogenic erosion
(B) Zoogenic erosion
(C) Phytogenic erosion
(D) Extraction

90. Sheet erosion is also termed as:
(A) Laminar erosion
(B) Phytogenic erosion
(C) Attrition (D) Detrition

91. Hydrologic processes of rainfall and run-off drive the:
(A) Erosion process.
(B) Sedimentation process
(C) Detachment of soil particles
(D) All of the above

92. The transportation of particles by direct splash to the rill is:
(A) Very large in amount
(B) About 50% of total
(C) Zero (D) Minimum

93. The sequence of water erosion is:
(A) Splash, sheet, rill and gully
(B) Sheet, rill, raindrop and gully erosion
(C) Rill, sheet, splash and gully erosion
(D) Gully, splash, rill and sheet erosion

94. Gully formation is mainly due to:
(A) land surface without vegetative cover and overgrazing
(B) Adoption of faulty tillage work
(C) Not checking of developed rills
(D) Improper construction of channels etc

95. Stage-3 gully development refers to:
(A) Healing stage
(B) Stabilizes stage
(C) Initiation stage
(D) None of the above

96. Relatively catchment area is larger for:
(A) V-shape gullies
(B) U-shaped gullies
(C) Inactive gullies
(D) none of the above

97. Relative flow velocity is higher in case of:
(A) V-shaped gullies
(B) Inactive gullies
(C) U-shaped gullies.
(D) None of the above

98. Bed width of medium gully is:
(A) More than 50 m
(B) Not less than 18 m
(C) 4 to 8 m
(D) 20 m

99. The flow velocity of run-off at which washing of soil particle takes place is Called as:
(A) Maximum velocity
(B) Critical velocity
(C) Permissible velocity
(D) Design velocity

100. Design of diversion ditches is based on return period of:
(A) 5 to 10 years (B) 50 years
(C) 25 years (D) 100 years

101. Location of permanent gully control structure is decided on the basis of:
(A) Gully depth (B) Gully width
(C) Gully bed slope (D) All of the above

102. The purpose of cut-off walls in drop structure is to:
(A) Provide structural strength against sliding
(B) Dissipate K E of flow
(C) Support gully walls
(D) Spill the flow safely

103. The apron length in rubble masonry dam should:
(A) Not less than 1.5 times the dam height
(B) Below 1.5 m
(C) Be equal to 1.5 times the dam height
(D) Be 3/4 of dam height

104. Permanent gully control structures are designed in respect of:
(A) Hydrologic
(B) Hydraulic
(C) Structure
(D) all of the above

105. Design of inlet section of straight drop structures is done by using:
(A) Weir formula
(B) Orifice formula
(C) Darcy's formula
(D) Flume formula

106. Hydraulic jump takes place, when flow enters from:
(A) Super critical to sub critical
(B) Critical state to super critical state
(C) Sub critical state to super critical state
(D) Uniform flow to non uniform flow

107. A hydraulic structure is expected to be safe against floating phenomenon, when resultant of horizontal and vertically forces is acting:
(A) Vertical downward
(B) Vertically upward
(C) Laterally
(D) None of the above

108. l30 Index method computes the rainfall erosivity factor based on:
(A) Rainfall intensity
(B) Rainfall duration
(C) Rainfall depth
(D) Run off depth

109. The average annual rainfall in India is about:
(A) 1194 mm (B) 1584 mm
(C) 1890 mm (D) 3000 mm

110. Contour bunds height is generally kept between:
(A) 2 to 3.0 m (B) 03 to 1.0 m
(C) 1.5 to 2.0 m (D) 5 to 10m

111. The construction of dugout type form pond is found suitable for area, where land slope is:
(A) Less than 4%
(B) 10%
(C) more than 15 %
(D) from 15 m depth

112. Soil loss phenomena is:
(A) A dynamic event
(B) A static event
(C) A cyclic event.
(D) None of the above

113. When drop height exceeds 4 m and there is possibility of silt accumulation in farm pond, then most suitable mechanical spillway to use is:
(A) Drop inlet spillway
(B) Straight drop spillway
(C) Grassed waterway
(D) Chute spillway

114. The most economical section of grass water way is:
(A) Parabolic
(B) Towards outlet
(C) Triangular
(D) Rectangular

115. GIS refers to:
(A) Geological information system
(B) Geographical information system
(C) Geometrical information system
(D) Geological information science

116. NRSA stands for:
(A) National Rural Space Administration
(B) National Remote Satellite Agency
(C) National Remote Space Agency
(D) National Remote Sensing Agency

Answers Key

Q. Ans	Q. Ans	Q. Ans	Q. Ans	Q. Ans	Q. Ans	Q. Ans	Q. Ans	Q. Ans
1. B	2. B	3. A	4. B	5. A	6. C	7. B	8. C	9. B
10. A	11. A	12. B	13. B	14. B	15. B	16. C	17. B	18. A
19. C	20. C	21. B	22. B	23. A	24. A	25. A	26. D	27. C
28. A	29. C	30. D	31. B	32. A	33. A	34. B	35. A	36. B
37. A	38. D	39. C	40. B	41. A	42. A	43. B	44. B	45. D
46. A	47. C	48. C	49. A	50. B	51. B	52. A	53. B	54. C
55. B	56. B	57. B	58. B	59. B	60. B	61. C	62. C	63. A
64. B	65. C	66. B	67. A	68. A	69. C	70. C	71. A	72. D
73. A	74. B	75. B	76. D	77. D	78. B	79. D	80. B	81. A
82. C	83. A	84. C	85. A	86. C	87. A	88. D	89. A	90. C
91. A	92. D	93. A	94. A	95. A	96. A	97. A	98. B	99. C
100. A	101. C	102. A	103. A	104. D	105. A	106. A	107. A	108. A
109. A	110. A	111. B	112. A	113. D	114. A	115. B	116. D	

CHAPTER 8

Asst. Engineer Comb. Exam (RPSC) & (Agricultural Egg.) Entrance Exam Questions

1. Levels and theodolites are usually equipped with two horizontal cross hairs which are called:
(A) Horiontal cross hairs
(B) Vertical cross hairs
(C) Stadia hairs
(D) Principal cross hairs

2. The volume of earthwork in land levelling can be computed by
(A) Volumetric method
(B) Prismoidal method
(C) End area method
(4)Arithmetic mean method

3. The ratio of the map distance to the corresponding distance of the ground is called as
(A) Scale
(B) Unit fraction
(C) Plan factor
(D) Representative factor

4. Which method of survey is the most accurate?
(A) Plane table method
(B) Grid method
(C) Transit method
(D) Chain survey method

5. The accuracy of measurement in chain surveying does not depend upon
(A) Length of the offset
(B) Scale of the plotting
(C) Importance of the features
(D) General layout of the chain lines

6. An ideal vertical curve to join two gradients is
(A) Circular (B) Parabolic
(C) Elliptical (D) Hyperbolic

7. If a 30m chain diverges by a perpendicular distance'd' from its correct alignment, the error in measured length is
(A) dx d/60 m (B) dx d/30 m
(C) dx d/40 m (D) d/30 m

8. The correct sequence of water erosion is
(A) Splash, Sheet, Rill, Gully
(B) Sheet, Gully, Rill, Splash
(D) Rill, Splash, Sheet, Gully
(D) Gully, Splash, Sheet, Rill

9. The thermal efficiency of diesel engine varies as
(A) 2-32% (B) 32-38%
(C) 38-42% (D) 42.52%

10. The line which separates the catchment basin from the rest of the area is
(A) Ridge line (B) Dum line
(C) Catchment line (D) Watershed line

11. A combination of contouring and crop rotation, in which alternate rows are constructed Perpendicular to the direction of wind and water flow, to prevent soil erosion, is called
(A) Strip cropping (B) Mulching
(C) Vegetated waterways
(D) Terracing

12. Water gas is a mixture of
(A) CO_2 and O_2
(B) O_2 and H_2
(C) H_2, N_2 and O_2
(D) CO, N_2 and H_2

13. What will be the height of an overflow spillway section for a design discharge of 1500m s, given that the upstream water surface level is al elevation of 240 m and the upstream channel floor is at 200 m; the spillway, having a vertical face, is 50 m long.
(A) 3.45 m (B) 4.55 m
(C) 5.00 m (D) 5.76 m

14. Which of the following does not constitute 90% of dry weight of any food?
(A) Carbohydrates (B) Fibres
(C) Proteins (D) Fats

15. Heating value of coal is approximately.
(A) 1000-2000 kcal/kg
(B) 2000-4000 kcal/kg
(C) 5000-6500 kcal/kg
(D) 9000-10,500 kcal/kg

16. In which decade wind-generated electricity was first sold to the public?
(A) 1940s (B) 1960s
(C) 1980s (D) 2000s

17. A meter suitable for total flow measurement is:
(A) Turbine flow meter
(B) Venturimeter
(C) Rotameter
(D) Orifice meter

18. The Water Horse Power (WHP) for centrifugal water pump for given flow rate of 4500 litre/sec and 10 m head will be:
(A) 14.26 (B) 16.26
(C) 18.26 (D) 12.26

19. Calculate the discharge of reciprocating pump (single acting) if area of cylinder is 0.25 m Length of stroke is 0.1 m, number of cylinder I and speed of pump is 50 rpm:
(A) 0.01125 m^3/s (B) 0.02125 m^3/s
(C) 0.03125 m^3/s (D) 0.04125 m^3/s

20. Pulverized coal is:
(A) Coal free from ash
(B) Non. smoking coal
(C) Coal which burns long time
(D) Coal broken into fine particles

21. Most of the solar radiation received on . earth's surface lies within the range of:
(A) 0.2 to 0.4 microns
(B) 0.38 to 0.78 microns
(C) 0 to 0.38 microns.
(D) 05 to 0.8 microns

22. A spillway has been designed for a head of 2.80 with length 200 m. The discharge Coefficient is 0.49. The discharge for this head will be:
(A) 2034 m^3/s (B) 2234 m^3/s
(C) 2434 m^3/s (D) 2634 m^3/s

23. In which collector air flows without any obstruction:
(A) Porous absorber plate.
(B) Non-porous absorber plate
(C) Over-lapped glass absorber
(D) Finned absorber

24. Which types of biogas plant are fed and emptied regularly?
(A) Batch type (B) Continuous type
(C) Dome type (D) Drum type

25. Reflecting mirrors used for harnessing solar energy are called:
(A) Mantle (B) Ponds
(C) Diffusers (D) Heliostats

26. The Mould Board plough absorbs side forces mainly through the:
(A) Share (B) Mould board
(C) Land side (D) Frog

27. The energy sources which produce no net energy are:
(A) Secondary energy sources
(B) Primary energy sources
(C) Commercial energy sources
(D) Non-commercial energy sources

28. **Heavy smoking in an engine during operation may be due to:**
    (A) Rich mixture (B) Overloading
    (C) Late injection (D) All of the above

29. **The voltage generated in spark plug at the time of spark is:**
    (A) 300-400 V (B) 100-200 V
    (C) 1000-5000 V (D) 20,000 V

30. **Disc angle of standard disc plough varies between:**
    (A) 15-25° (B) 25-35°
    (C) 35-42° (D) 42-45°

31. **When an incandescent light bulb is turned on, what percent of the electricity is converted into light energy?**
    (A) 10% (B) 30%
    (C) 50% (4)70%

32. **What percent of the average home's electric bill is for lighting?**
    (A) About 10% (B) About 20%
    (C) About 30% (D) About 40%

33. **Which one of the following pair is not correctly matched?**
    (A) Solidity-Wind Machine
    (B) Anaerobic digestion of Organic matter-Bio gas generation
    (C) Destructive dilation- Charcoal Production
    (D) Complete Combustion- Producer gas

34. **Which one of the following statement is correct?**
    (A) Surge irrigation is used along with centre pivot swing arm to provide more uniform irrigation
    (B) The lateral carries water from pump to mainline
    (C) Organic matter percentage and soil texture affect infiltration rate of water
    (D) If water is applied to soil in excess of infiltration rate even runoff will not occur

35. **From the hydraulic efficiency point of view, the most efficient cross-section of an open channel is:**
    (A) Parabolic (B) Trapezoidal
    (C) Semi-circular (D) Rectangular

36. **Watershed is a**
    (A) Hydrological entity contributing runoff to a common point
    (B) Hydrological entity receiving runoff to a single point
    (C) Hydrological entity receiving runoff from multi. inlet
    (D) Hydrological entity distributing runoff from common inlet

37. **The depth of flow over a sharp crested rectangular weir should not be more than about:**
    (A) Half the crest width
    (B) One third of the crest width
    (C) Three-fourth of the crest width
    (D)The width of the weir

38. **The term Tons of refrigeration is used in cold storage unit to measure.**
    (A) Mass of fruits held
    (B) Mass of refrigerant
    (C) Mass of compressor
    (D) Capacity of compressor

39. **Which of the following instrument is used for measurement of specific heat?**
    (A) Micrometer (B) Calorimeter
    (C) Thermometer (D) Current meter

40. **The ratio of force of inertia and friction force in a flow may be characterized by**
    (A) Nusselt number
    (B) Reynolds number
    (C) Rayleigh number
    (D) Peclet number

41. **The mass of water vapour per unit mass of dry air is**
    (A) Specific humidity
    (B) Percentage humidity
    (C) Relative humidity
    (D) Equilibrium relative humidity

42. **If the number of microbes in a process has to be reduced from a total load of 106 to a final 104, the required thermal death time (D Value) will be.**

(A) 10D (B) 4D
(C) 5D (D) 20D

43. Water horse power of centrifugal pump of 15 litres/sec capacity and 30m total head will be equal to
(A) 3.0 (B) 4.0
(C) 5.0 (D) 6.0

44. Interceptor drain helps to control water logging by
(A) Lowering the water table
(B) Preventing subsoil water from reaching the area
(C) Allowing vertical drainage
(D) Darning out excess al the natural drain

45. The actual area irrigated in a year from an outlet is
(A) Irrigation period
(B) Intensity of irrigation
(C) Irrigation frequency
(D) Irrigation efficiency

46. The minimum side slope of an earth channel for polyethylene lining is
(A) 1.5:1 (B) 2:1
(C) 2.5:1 (D) 3:1

47. Which type of cultivating tool is extensively used for weed control?
(A) Furrower type
(B) Sweep type
(C) Slip nose type
(D) Double point shovel type

48. A wheel of an implement to maintain a uniform depth of working in soil is known as.
(A) Rear furrow wheel
(B) Land wheel
(C) Front furrow wheel
(D) Gauge wheel

49. For the farm area of 40 ha land under single cropping pattern, the suitable size of tractor.
(A) 10 to 20 HP
(B) 50 HP
(C) 20 to 25 HP
(D) 35 HP

50. The ratio of total force output of the traction device in the direction of travel to the dynamic weight on the traction device is called.
(A) Rolling resistance
(B) Coefficient of traction
(C) Tractive efficiency
(D) Coefficient of friction

51. When the speed of a centrifugal pump is changed the head varies as
(A) The speed
(B) Square of the speed
(C) Square root of the speed
(D) Cube of the speed

52. The Cetane rating of commercial diesel fuel is between
(A) 1 to 30 (B) 60 to 90
(C) 30 to 60 (D) 90 to 120

53. Ditch conduit formula is used for
(A) Sand bearing test.
(B) Size of tile drain
(C) Grade of tile drain.
(D) Loads on drain pipes

54. Cavitation is formation of cavity in flow due to
(A) Negative pressure as a result of high velocity
(B) Positive pressure as a result of high velocity
(C) Reduction in pressure as a result of high velocity of water
(D) None of the above

55. What is the side draft of a tillage implement?
(A) Horizontal component of pull, parallel to the line of motion
(B) Vertical component of pull
(C) Horizontal component of pull perpendicular to the line of motion
(D) Pull divided by the furrow cross section area

56. An aquifer which is bounded by an impermeable layer at the bottom and semi-pervious layer at the top is known as

(A) Artesian aquifer
(B) Confined aquifer
(C) Semi-confined aquifer
(D) Unconfined aquifer

57. **Cutter bar is a component of**
(A) Seed drill
(B) Flail mower
(C) Vertical conveyor reaper
(D) Hand hoe

58. **Furrow cross-section in a country plough is commonly.**
(A) Circular (B) Parabolic
(C) Rectangular (D) Trapezoidal

59. **Spiked tooth type threshing cylinder are generally used in**
(A) Rice threshers
(B) Wheat threshers
(C) Maize threshers
(D) Mustard threshers

60. **The crank shaft and rear axle of tractor are attached at an angle of**
(A) 30° (B) 60°
(C) 90° (D) 120°

61. **The drainage density is**
(A) Length of streams per drainage area within basin
(B) Stream discharge per unit drainage area
(C) Drainage area per unit stream length
(D) Number of streams per unit drainage area

62. **Value of Drainage coefficient for small drainage projects is taken as.**
(A) 10 to 15 mm/day
(B) 5 to 10 mm/day
(C) 6 to 25 mm/day
(D) 15 to 20 mm/day

63. **The ratio of the partial pressure of moisture of the food to vapour pressure of the water at the same temperature is called.**
(A) Moisture content
(B) Water activity
(C) Dry bulb temperature
(D) Wet bulb temperature

64. **Methane $(CH)_4$ content in producer gas obtained from biomass gasification is**
(A) 60% (B) 1.5%
(C) 6-10% (D) 11-20%

65. **High solidity wind rotor works on**
(A) Drag force
(B) Lift force
(C) Both Lift and Drag force
(D) Gravity force

66. **Calorific value of biogas (60% CH4) is**
(A) 5713 kcal/kg
(B) 4713 kcal/m^3
(C) 4200 kcal/m^3
(D) 3713 kcal/m^3

67. **Which one among these will be the most suitable dryer for drying liquid foods?**
(A) Dielectric dryer
(B) Spin flash dryer
(C) Tunnel dryer
(D) Trough dryer

68. **The crawler tractor have .... type of brake.**
(A) Hydraulic
(B) Internal expanding shoe
(C) Disc
(D) External contracting shoe

69. **The problem of "back logging" occurs in**
(A) Belt conveyor
(B) Screw conveyor
(C) Bucket elevator
(D) Pneumatic conveyor

70. **The term mesh" refers to the number of openings per square**
(A) Decimeter (B) Millimeter
(C) Centimeter (D) Inch

71. **The type of the induction motors having variable speed characteristics is**
(A) Split phase (B) Capacitor
(C) Universal (D) Three phase

72. **Water mist polishing machine is used for the milling of**
(A) Wheat (B) Maize
(C) Rice (D) Barley

73. Which one is the most suitable method of drying for fruit juice?
(A) Solar drying
(B) Fluidized bed drying
(C) Tray drying
(D) Spray drying

74. Table separators work on the basis of
(A) Size
(B) Shape
(C) Specific gravity
(D) Terminal velocity

75. Which of the erosion is extremely harmful for the land?
(A) Raindrop erosion
(B) Sheet erosion
(C) Rill erosion
(D) Gully erosion

76. The glazing which limits the radiation and convection heat losses is
(A) Absorber plate
(B) Selective surface
(C) Insulation
(D) Transparent cover

77. In a hydro-electric plant, spillways are used
(A) To discharge all surplus water
(B) To discharge surplus water on the downstream side of dam
(C) When water is not available in sufficient quantity
(D) None of the above

78. Which one of the following pairs is not correctly matched?
(A) Well log–The practice of making a detailed record of the geologic formations penetrated by a borehole.
(B) Propeller–A type of fan that transmits power by converting rotational motion into thrust.
(C) Turbine pump–Pump used for electricity generation
(D) Sprinkler irrigation–Modern irrigation technique

79. The commonly used square sizes for contouring is
(A) 10m × 10m to5m × 5m
(B) 10m × Om to 15m × 15m
(C) 5m × 5m to 20 × 20m
(D) 5m × 5m to 10m × 15m

80. The length of a line measured with 20m chain was found to be 500m. It was subsequently found that the chain was longer by 0.04m. What is the correct length of line?
(A) 500m
(B) 501 m
(C) 502 m
(D) 503 m

81. The equation for curve number is given as
(A) CN =2540/25.4 + S
(B) CN= 25.4 + S/2540
(C) CN=2500/25.0+S
(D) CN=2540 +S/25.4

82. The main function of crank shaft in an engine is
(A) To convert reciprocating motion into rotary motion
(B) To operate engine valves
(C) To operate fuel injection pump
(D) To increase engine efficiency

83. Value of Betz's coefficient is
(A) 0.562
(B) 0.593
(C) 0.621
(D) 0.657

84. The time of retardation case of creep test is
(A) 36.7% of the time for total strain
(B) 63.2% of the time for total strain
(C) 76.3% of the time for total strain
(D) 83.2% of the time for total strain

85. Thermostat valve is a component of
(A) Engine fuel system
(B) Exhaust system
(C) Cooling system
(D) Hydraulic system

86. The ratio between the molecular diffusivity of momentum to the molecular diffusivity of heat is
(A) Nusselt number (B) Prandtl number
(C) Rayleigh number (D) Peclet number

87. The time of concentration of a watershed is proportional to
(A) $L1^{-77}$ (B) $S^{-0.385}$
(C) $L1^{-77} S^{0.385}$ (D) $S1^{-77}$

88. Hydrographic Survey deals with the mapping of
(A) Large water bodies
(B) Watershed
(C) Canal system
(D) Movements of clouds

89. The grazing animals can change the type of Vegetation by
(A) Cross pollination.
(B) Selective grazing
(C) Bringing in plant pathogens
(D) Bringing the seeds of other plants

90. Contour bunding is done to check.
(A) Sheet erosion
(B) Rill erosion
(C) Gully erosion
(D) Ravine formation

91. Interception loss is
(A) High in the beginning of storm and gradually decreases
(B) Low in the beginning of storm and gradually decreases
(C) High in the ending of storm and gradually decreases
(D) High in the beginning of storm and gradually increases

92. The heat transfer rate in solid agricultural products of any shape is called
(A) Mass density (B) True density
(C) Bulk density (D) Mass diffusivity

93. Octane number for regular petrol fuel varies as
(A) 95-98 (B) 8-90
(C) 70-80 (D) 60-65

94. Critical pressure of water is
(A) 1 kg/cm^2 (B) 100 kg/cm^2
(C) 155 kg/cm^2 (D) 213.8 kg/cm^2

95. Which of the following countries is not in the list of top 3 consumers of energy in the world?
(A) China (B) India
(C) Russia (D) United States

96. Which one is a false statement?
(A) Magnets are used to generate electricity
(B) Using hydropower does not impact the environment
(C) Propane comes from oil and natural gas wells
(D) Electricity is a renewable resource

97. The maximum size of the raindrop is
(A) 6 mm (B) 10 mm
(C) 2 mm (D) 0.5 mm

98. In the Universal Soil Loss Equation (USLE), the soil erodibility factor K is ....
(A) Measure of the susceptibility of soil particles to detachment and transport
(B) Slope length gradient factor
(C) Crop management factor
(D) Rainfall-runoff factor

Answers Key

Q. Ans	Q. Ans	Q. Ans	Q. Ans	Q. Ans	Q. Ans	Q. Ans	Q. Ans	Q. Ans
1. C	2. B	3. D	4. B	5. D	6. B	7. A	8. A	9. B
10. A	11. A	12. D	13. D	14. B	15. C	16. A	17. A	18. D
19. C	20. D	21. A	22. A	23. B	24. B	25. D	26. C	27. A
28. D	29. D	30. D	31. A	32. *	33. D	34. C	35. C	36. A
37. B	38. D	39. B	40. B	41. A	42. A	43. D	44. B	45. B
46. *	47. B	48. D	49. C	50. B	51. B	52. C	53. D	54. C
55. C	56. C	57. C	58. D	59. A	60. C	61. A	62. C	63. B
64. B	65. A	66. B	67. B	68. D	69. C	70. D	71. C	72. C
73. D	74. C	75. B	76. D	77. B	78. C	79. C	80. B	81. A
82. A	83. B	84. B	85. C	86. B	87. B	88. A	89. B	90. A
91. A	92. D	93. B	94. D	95. B	96. D	97. A	98. A	

CHAPTER 9

Ph.D. ENTRANCE EXAM QUESTIONS

1. Which of the following crops have been approved for commercial cultivation in India?
(A) *Bt* cotton and *Bt* brinjal
(B) *Bt* cotton and Golden Rice
(C) *Bt* maize and *Bt* cotton
(D) *Bt* cotton only

2. For the year (2010-11) the expected food grain production in India was
(A) 212 million tonnes
(B) 220 million tonnes
(C) 235 million tonnes
(D) 250 million tonnes

3. The genome of which of the following crops is still not completely sequenced?
(A) Rice (B) Soybean
(C) Sorghum (D) Wheat

4. Which of the following sub-schemes are not covered under the Rashtriya Krishi Vikas Yojana?
(A) Extending the Green Revolution to eastern India
(B) Development of 60,000 pulses and oilseeds villages in identified watersheds
(C) National Mission on Saffron
(D) National Mission on Bamboo

5. According to the Human Development Report 2010 of the United Nations, India's rank in terms of the human development
(A) 119 (B) 134
(C) 169 (D) 182

6. Which of the following does not apply to SRI method of paddy cultivation?
(A) Reduced water application
(B) Reduced plant density
(C) Increased application of chemical fertilizers
(D) Reduced age of seedlings

7. Which organic acid, often used as a preservative, occurs naturally in cranberries, prunes, cinnamon and cloves?
(A) Citric acid (B) Benzoic acid
(C) Tartaric acid (D) Lactlc acid

8. Cotton belongs to the family
(A) Cruciferae (B) Anacardiaceae
(C) Malvaceae (D) Solanaceae

9. Photoperiodism is
(A) ending of shoot towards source of light
(B) Effect of light/dark durations on physlological processes
(C) Movement of chloroplast in cell in response to light
(D) Effect of light on chlorophyll synthesis

10. Ergot disease is caused by which pathogen on which host?
(A) Claviceps purpurea on rye
(B) Puccinia recondita on wheat
(C) Drechlera sorokiniana on wheat
(D) Albugo candida on mustard

11. Rocks are the chief sources of parent materials over which soils are developed Granite, an important rock, is classified as
(A) Igneous rock
(B) Metamorphic rock
(C) Sedimentary rock
(D) Hybrid rock

12. Which one of the following is a Kharif crop?
(A) Pearl millet (B) Lentil
(C) Mustard (D) Wheat

13. The coefficient of variation (C.V.) is calculated by the formula
(A) (Mean/S.D.) × 100
(B) (S.D./Mean) × 100
(C) S.D./Mean
(D) Mean/S.D

14. Which of the following is commonly referred to as murate of potash?
(A) Potassium nitrate
(B) Potassium chloride
(C) Potassium sulphate
(D) Potassium silicate

15. For applying 100 kg of nitrogen, how much urea would one use?
(A) 45 kg (B) 111 kg
(C) 222 kg (D) 333 kg

16. The devastating impact of plant disease on human suffering and survival was first realized by epidemic of
(A) Brown spot of rice in Bengal
(B) Late blight of potato in USA
(C) Late blight of potato in Europe
(D) Rust of wheat In India

17. The species of rice (Oryza) other than O. sativa that is cultivated is
(A) O. rufipugon
(B) O. fongisteminata
(C) O. glaberrima
(D) O. nivara

18. The enzyme responsible for the fixation of CO_2 in mesophyll cells of C-4 plants is
(A) Malic enzyme
(B) Phosphoenol pyruvate carboxylase
(C) Phosphoenol pyruvate carboxykinase
(D) RuBP carboxylase

19. Which one of the following is a Vertisoil?
(A) Black cotton soil
(B) Red sandy loam soil
(C) Sandy loam sodlc soil
(D) Submontane (Tarai) soil

20. What is the most visible physical characteristic of cells in metaphase?
(A) Elongated chromosomes
(B) Nucleus vislble but chromosomes not
(C) Fragile double stranded loose chromosomes
(D) Condensed paîred chromosomes on the cell plate

21. All weather phernomena like rain, fog and mist occur in
(A) Troposphere (B) Mesosphere
(C) lonosphere (D) Ozonosphere

22. Which of the following elements is common to all proteins and nucleic acids?
(A) Sulphur (B) Magnesium
(C) Nitrogen (D) Phosphorous

23. Silt has intermediate characteristics between
(A) Sand and loam
(B) Clay and loam
(C) Loam and gravel
(D) Sand and clay

24. Certified seed is produced from
(A) Nucleus seed
(B) Breeder seed
(C) Foundation seed
(D) Truthful seed

25. Seedless banana is an:
(A) Auto triploid
(B) Auto tetraploid
(C) Allo triploid
(D) All otetraploid

26. Which one of the following is used to test the goodness-of-fit of a distribution?
(A) Normal test
(B) t-test
(C) chi-square test
(D) F-test

27. The Subsurface runoff is also known as
(A) Interflow
(B) Storm seepage
(C) Secondary base flow
(D) All of the above

28. Soil structure refers to which of the following?
(A) Arrangements of soil particles
(B) Proportion of sand, silt, clay
(C) Organic composition
(D) Profile depth

29. Why is sediment yield often less than soil erosion in a given drainage basin?
(A) Because streams usually cannot carry all of the sediment that is eroded
(B) Because sediment yield is totally unrelated to soil erosion
(C) Because streams naturally evolve towards a graded condition
(D) Because a stream's bed load usually exceeds its suspended load

30. The volume of water passing by per unit time is called the volume flow rate. What happens to the volume flow rate of the water as it passes from thick section to the tin section?
(A) It will decrease (B) Will increase
(C) It won't change (D) Cannot know

31. Isotherm is a line which joins the points of equal
(A) Rainfall depth
(B) Temperature
(C) Humidity
(D) Atmospheric pressure

32. Soil having highest infiltration capacity
(A) Sandy clay loam (B) Clay loam
(C) Loam (D) Silty clay

33. During which of the following does water move in a direction different from others?
(A) Percolation (B) Transpiration
(C) Infiltration (D) Precipitation

34. Climate is influenced by
(A) The amount of incoming solar radiation
(B) Earth's rotation
(C) The tilt of earth's axis
(D) The moon's gravity

35. Which of the following statements is false?
(A) Recharging of water is a slow process
(B) The water table moves down
(C) Water in a confined aquifer is under pressure
(D) Groundwater is stationary and does not move

36. The main difference between a suspension and a colloid is that
(A) In suspension, the particles eventually settle to the bottom
(B) In suspension, the solute is permanently dissolved in the solvent
(C) In colloids, the particles eventually settle to the bottom
(D) In colloids, the solute is permanently solvent

37. In the case of a water table well, the piezometer surface
(A) As above the ground level
(B) Is below the water level in the well
(C) Between the ground .level and the water surface in the well
(D) Below the water surface in the well

38. Which of the following has the highest infiltration capacity?
(A) Forest land
(B) Grazed pasture
(C) Concrete pavement in airport
(D) Rocky land

39. The salinity in water
(A) Reduces the evaporation
(B) Does not affect evaporation
(C) Increases the evaporation
(D) Difficult to say

40. The permeability of an aquifer
(A) Increases with increase in temperature
(B) Increases with decrease in temperature
(C) is independent of temperature
(D) Decreases with decrease in temperature

41. How is water stored in an aquifer?
(A) In an open underground lake
(B) In cracks and spaces, in rocks
(C) In impermeable rock
(D) In wells and springs

42. Evaporation through plants and surrounding soil together is called
(A) Evaporation
(B) Transpiration
(C) Evapotranspiration
(D) All of the above

43. The problem of flood routing through a reservoir may be approached using
(A) The law of, conservation of mass
(B) Newton's second law
(C) The law of conservation of energy
(D) Darcy's law

44. How much of earth's water is fresh water?
(A) 97% (B) 50%
(B) 3% (D) 1%

45. What is the best definition of georeferencing?
(A) The process of assigning coordinates from a known reference system to spatial data
(B) The process of matching address locations with street data
(C) The process of assigning datum and spheroid to the data
(D) The process of transferring geographic coordinates from one projection to another

46. What will happen to the velocity of water as it passes from a relatively thick section of pipe to a relatively thin section of pipe?
(A) It will increase
(B) It will slow down
(C) The velocity won't change
(D) Can't say

47. For a given storm, the (average depth of rainfall over an area)
(A) Decreases with increase in area
(B) Increases with Increase in area
(C) Has no relation with area
(D) Decreases with decrease in area

48. The average equivalent ratio of (Ca+Mg) to HCO_3 is
(A) Lower in the high-land river waters than that in the low-land river waters
(B) Higher in the high-land river waters than that in the low-land river waters
(C) Double in the low-land river waters as compared with that in the high-land river waters
(D) More than double in the low-land river waters as compared with that in the high-land river waters

49. Based on PET, available water holding capacity of soils and both moisture and the thermal classifications, the climate of India has been divided into how many zones?
(A) 10 (B) 8
(C) 6 (D) 4

50. The areas covering northern Punjab, Haryana, Uttar Pradesh, Bihar, West Bengal, Orissa, Madhya Pradesh and northern part of Andhra Pradesh fall under
(A) Humid climate
(B) Sub-humid climate
(C) Pre-humid climate
(D) Semi sub-humid climate

51. The cereal production on account of climate change is likely to decline
(A) Everywhere in the world
(B) Only in the developed countries
(C) Only in the developing countries
(D) Both in the developed and developing countries

52. Meteorological drought in the area is defined as a situation where the actual rainfall decreases by more than
(A) 40 percent of the normal rainfall
(B) 30 percent of the normal rainfall
(C) 25 percent of the normal rainfall
(D) Z0 percent of the normal rainfall

53. **The evaporation from the semi-dry soil, which occurs 1-5 days after irrigation, depends on**
    (A) Climatic factors
    (B) Type of the soil
    (C) Characteristics of the crop
    (D) Hydraulic conductivity of the soil

54. **The peak period evapotranspiration is required for**
    (A) Determining irrigation water supply
    (B) Evaluating the performance of an irrigation project
    (C) Planning an irrigation system
    (D) Developing irrigation schedules

55. **The major role of water user's association in a canal irrigated command at present is to**
    (A) Decide on season's water allocation
    (B) Consult and look to the irrigations departments for water supply
    (C) Ensure that outlets and structures are not tempered with
    (D) Distribute water below outlets

56. **The efficiency of Decision Support System (DSS) is the**
    (A) Measure of the skill of the decision makeing
    (B) Degree to which goals are achieved
    (C) Success in developing the linkages between the objects
    (D) Measure of the use of inputs

57. **The main difficulty in developing an expert system for planning and operating of water resource systems is associated with the**
    (A) Effective manipulation of large Knowledge bases
    (B) Effective manipulation of large databases
    (C) Complexities of the changes on the programmers
    (D) Lack of availability of input data

58. **Remote sensing may be used to carry out ground water investigations by performing**
    (A) Geohydromorphological maps
    (B) Hydrogeomorphological maps
    (C) Geomorphologic maps
    (D) Gephydrological maps

59. **Major irrigation projects includes**
    (A) Only surface water based schemes irrigating more than 10,000 ha area
    (B) Only surface water based schemes irrigating more than 20,000 ha area
    (C) Both the surface and ground water based schemes irrigating more than 10,000 ha area
    (D) Both the surface and ground based schemes irrigating more than 20,000 ha area

60. **The major reason for the gap between the created and the utilized irrigation potential under the medium and major Irrigation projects as reported by the Committee set up by the Ministry of Irrigation and Power in 1972 was**
    (A) Non-availability of field channels
    (B) Use of unreliable, data for designing and constructing irrigation network
    (C) Excessive seepage losses in the canal network
    (D) Absence of water users' associations

61. **One of the several limitations of the Kastiakov's equation used for determining infiltration rate of water into soil is that it does not fit well for**
    (A) Small values of elapsed time
    (B) Large values of elapsed time
    (C) Coarse-textured soil
    (D) Fine-textured sol

62. **If the depth of water in an earthen channel is increased, the percent seepage loss in the channel will be**
    (A) Proportionally with the increase in. the carrying capacity of the channel
    (B) Proportionally less than the increase in the depth
    (C) Proportionally more than the increase in the depth
    (D) Proportionally with the increase in the depth

63. The applicability of the Darcy's law in any soil
(A) Depends upon soil texture
(B) Depends on soil structure
(C) Depends both on soil texture and soil structure
(D) Is independent of soil texture and soil structure

64. The total flow rate of water in a saturated soil is proportional to the
(A) Radius of the pores
(B) Square of the radius of the pores
(C) Third power of the radius of the pores
(D) Fourth power of the radius of the pores

65. The initial weight of a 2×2 m Lysimeter after irrigating a crop was measured as 6.35 tonnes and its weight 10 day after irrigation was found to be 6.15 tonnes. If the volume of the drainage water collected was 100 litres the average value of the daily evapotranspiration of the crop will be
(A) 2.75 mm (B) 2.50 mm
(C) 2.25 mm (D) 2.00 mm

66. A water sample with a $CaCO_3$ concentration of 150 mg/litre may be classified as
(A) Soft
(B) Moderately soft
(C) Hard
(D) Moderately hard

67. The spacing between tiles for sub-surface drainage in a silty-clay loam is
(A) Less than that in a clay soil
(B) Greater than that in a clay-loam
(C) Greater than that in a sandy-loam
(D) Less than that in a clay. loam

68. If the field capacity of a soil is 30 percent, moisture before irrigation is 20 percent, apparent specific gravity of the soil is 1.6 root zone depth 1 m and daily consumptive use is 8 m, the irrigation interval is
(A) 25 days (B) by 20 days
(C) 15 days (D) 10 days

69. The water availability to the plants in a soil
(A) Remains the same over the entire range of water availability
(B) Depends upon the water table level depth
(C) Remains the same up to the critical soil moisture level
(D) Decreases with the water potential

70. If surface irrigation system is designed such that the advance phase of water front takes only 25 percent of the total Irrigation time in a soil, the infiltration rate of which is inversely proportional to the square root of the elapsed time, the water application efficiency of the system can be expected to be more than
(A) 90 percent (B) 80 percent
(C) 70 percent (D) 60 percent

71. The design and management studies in surface irrigation systems are carried out to increase
(A) Water use efficiency
(B) Application efficiency
(C) Storage efficiency
(D) Distribution efficiency

72. In a perfectly designed check basin system of irrigation
(A) Recession phase is more important than the depletion phase
(B) Depletion phase is more important than the recession phase
(C) Either of the two phases are equally important
(D) None of the two phases are important

73. Water storage efficiency is the ratio between
(A) Water stored in the root zone and the water applied to the field
(B) Water used by the plant and water stored in the root zone
(C) Water stored in the root zone and the water holding capacity of the root zone
(D) Water released from the root zone and water stored in the root zone

74. The friction loss per unit length in a lateral pipe with the same size as that of the main pipe of a pressurized irrigation system
(A) Is more than that in the main pipe
(B) Remains the same as that in the main pipe
(C) Increases with the number of the cutlets in the lateral pipe
(D) Is less than that in the main pipe

75. The groundwater flow into a well is governed by a Dupuit assumption which states that the flow velocity is proportional to the
(A) Sine of the hydraulic gradient
(B) Cosine of the hydraulic gradient
(C) Tangent of the hydraulic gradient
(D) Cotangent of the hydraulic gradient

76. Out of the four surface geophysical methods, the most widely applied method for groundwater investigation is the
(A) Gravity method
(B) Seismic reflection method
(C) Seismic refraction method
(D) Resistivity method

77. One of the most widely accepted soil and water conservation practices for watersheds with slopping agricultural lands is
(A) Contour and graded bunding
(B) Buffer strip cropping
(C) Contour cultivation
(D) Land levelling

78. The moisture content in a field has to be raised from 12 percent to 20 percent through irrigation. If the apparent specific gravity of the soil is 1.5 and the root zone depth is 50 cm, the depth of water to be added would be
(A) 5 cm (B) 6 cm
(C) 7 cm (D) 8 cm

79. The velocity of a particle travelling with water through a sediment basin is inversely proportional to the
(A) Flow rate
(B) Length of the basin
(C) Depth of the basin
(D) Concentration of the sediment

80. Hydraulic ram is an impulse pump and is specially suitable for
(A) Coastal areas (B) Humid areas
(C) Hilly areas (D) Alluvial areas

81. Water hammer occurs in a pipe when the flowing fluid expresses a sudden reduction
(A) Pressure (B) Pipe diameter
(C) Velocity (D) Pipe roughness

82. Isohyets are drawn by
(A) Joining rain gauge stations
(B) Drawing lines of equal rainfall for a given duration
(C) Orthogonal lines joining rain gauge stations
(D) Lines of equal elevations

83. Flood routing is the computation of Storage, when the outflow hydrograph is known
(A) Direction of the flood flow
(B) Inflow hydrograph, when the outflow hydrograph is known
(C) Outflow hydrograph, when the inflow
(D) Hydrograph and storage properties are known

84. Flow in an irrigation channel is assumed as
(A) Gradually varied (B) Spatially varied
(C) Uniform (D) Rapidly varied

85. The velocity distribution across a pipe flowing full is
(A) Triangular (B) Parabolic
(C) Uniform (D) Semi-circular

86. A practical method of determination of crop ETs through experimentation on
(A) Small pots
(B) Field plot
(C) Non weighing lysimeter
(D) Weighing lysimeter

87. Saline lands are mostly found in the region which are
(A) Temperate
(B) Arid and semi-arid
(C) Unirrigated
(D) Humid

88. In ground water recharge tube well, the recharge rate
(A) Decreases with time
(B) Increases with time
(C) Remains constant
(D) Fluctuates around a mean

89. Centrifugal pumps are used when suction lift is
(A) Less than 9 m
(B) Equal to 15 m
(C) Equal to 20 m
(D) Greater than 20 m

90. An average discharge of a pump 230 liters g per minute irrigates 100 m × 100 m wheat crop in 2 days and 2 hours, then what is the average depth of irrigation?
(A) 0.030 m
(B) 0.040 m
(C) 0.049 m
(D) 0.069 m

91. Water year in India starts from the first
(A) January
(B) March
(C) April
(D) June

92. A low drainage density of an area indicates
(A) High permeability of the terrain
(B) Low permeability of the terrain
(C) Dense network of streams
(D) Number of first order streams are more

93. A canal is 80km long with an average surface width of 15-m˙ Class A pan evaporation is 0.5 cm/d, volume of water evaporated in a month of 30 days in m3 is
(A) 1,75,000 (B) 1,26,000
(C) 18,000 (D) 12,600

94. What size of electric motor is required for pumping 300 L/min against a head of 50 m assuming a pump efficiency of 65%?
(A) 7.5 hp (B) 5hp
(C) 3 hp (D) 2 hp

95. The concept used to evaluate the sprinkler irrigation system is
(A) Uniform intensity
(B) Uniformity coefficient
(C) Coefficient of variation
(D) Uniformity index

96. Hydraulically the most efficient irrigation channel section is
(A) Semi-circular
(B) Trapezoidal
(C) Rectangular
(D) Elliptical

97. The commonly used radioactive isotope is groundwater development is
(A) Uranium (B) Thorium
(C) Tritium (D) Germanium

98. All watershed development works should start from
(A) Lower reaches of the watershed
(B) Middle reaches of the watershed
(C) Upper reaches of the watershed
(D) From anywhere in the watershed

99. Which of the following designs you would adopt for an experiment with. Soil salinity levels as main treatment and drip irrigation amount levels as sub treatment?
(A) Completely randomized design
(B) Randomized block design
(C) Factorial randomized block design
(D) Split plot design

100. For assessment of soil moisture in taking up rabi sowing in a dryland area of large extent, which of the following remote sensing data you would use?
(A) Far Infrared (FIR)
(B) Near infrared (NIR)
(C) Microwave
(D) Visible region

101. In rational method for computing the peak rate of runoff of watersheds, the factor implies the
   (A) Intensity of rainfall for 30 min duration
   (B) Intensity of rainfall for 1 hour duration
   (C) Intensity of rainfall for the duration equal to the time of concentration
   (D) Intensity of rainfall for the duration equal to half the time of concentration

Answers Key

Q.	Ans	Q.	Ans	Q.	Ans	Q.	Ans	Q.	Ans	Q.	Ans	Q.	Ans	Q.	Ans	Q.	Ans
1.	D	2.	C	3.	A	4.	A	5.	A	6.	C	7.	B	8.	C	9.	A
10.	A	11.	A	12.	A	13.	B	14.	B	15.	C	16.	A	17.	C	18.	B
19.	A	20.	A	21.	A	22.	C	23.	D	24.	C	25.	A	26.	C	27.	D
28.	A	29.	B	30.	B	31.	B	32.	C	33.	B	34.	A	35.	D	36.	A
37.	C	38.	A	39.	D	40.	A	41.	B	42.	C	43.	A	44.	C	45.	A
46.	A	47.	C	48.	B	49.	C	50.	B	51.	C	52.	C	53.	D	54.	C
55.	D	56.	A	57.	A	58.	D	59.	A	60.	A	61.	B	62.	B	63.	A
64.	D	65.	B	66.	C	67.	C	68.	D	69.	C	70.	D	71.	A	72.	A
73.	D	74.	A	75.	A	76.	B	77.	B	78.	A	79.	D	80.	B	81.	C
82.	B	83.	C	84.	B	85.	B	86.	B	87.	B	88.	A	89.	A	90.	D
91.	D	92.	A	93.	C	94.	A	95.	B	96.	A	97.	C	98.	C	99.	B
100.	D	101.	C														

CHAPTER 10

MATCH THE FOLLOWING (MATCHING)

S. No.	Group I		Group II
1.	Groundwater dating	(A)	Ground water movement
2.	Rational method	(B)	Plant growth
3.	Thiessen polygon	(C)	Carbon.14
4.	Well hydraulics	(D)	Rainfall
5.	Crop coefficient	(E)	Flood

Answer: 1-C, 2-E , 3-D, 4-A, 5-B

S. No.	Group I		Group II
1.	Hyetograph	(A)	Infiltration
2.	Capillary zone	(B)	Remote sensing
3.	Field capacity	(C)	Rainfall
4.	Land use/ Land cover	(D)	Irrigation
5.	Duty	(E)	Above water table

Answer: 1-C, 2-E, 3-A, 4-B, 5-D

S. No.	Group I		Group II
1.	Infrared	(A)	Agronomical measure
2.	Contour bunding	(B)	Drought
3.	Reverse osmosis	(C)	Evaporation
4.	Aridity index	(D)	Water quality
5.	Lysimeter	(E)	Surface water body

Answer: 1-E, 2-A, 3-D, 4-B, 5-C

S. No.	Group I		Group II
1.	Contour farming	(A)	To prevent sheet and ril erosion
2.	Strip farming	(B)	To perform soll manipulation
3.	Conservation tillage	(C)	To provide the ponding of water
4.	Contour trenching	(D)	To filter soil out of rain water
5.	Terracing	(E)	To held water and prevent

Answer: 1-D, 2-E, 3-B, 4-C, 5-A

Matching (S & WC Engg.)

S. No.	Group I		Group II
1.	Area velocity method	(A)	Bubbler
2.	Ephemeral stream	(B)	Herring bone
3.	Effluent stream	(C)	Booster pump
4.	Lysinmeter	(D)	Evapotranspiration
5.	Tile drain	(E)	Sewage water
6.	Ring basin irrigation	(F)	Ramp-cum-waste weir
7.	Sprinkler irrigation	(G)	Mole plough
8.	Subsurface drainage	(H)	Mean velocity
9.	Contour bund	(I)	Peak flood where no gauging station exists
10.	Semi-open pump	(J)	Perennial stream

Answers : 1-H, 2-J, 3-I, 4-D, 5-B, 6-A, 7-E, 8-G, 9-F, 10-C

S. No.	Group I		Group II
1.	Slip circle method	(A)	Stability of slopes
2.	Blanket	(B)	Earth dams
3.	Tile drain	(C)	Bench terrace
4.	Hooghoudt	(D)	Tube well
5.	Soil capability classification	(E)	Watershed management
6.	Lift irrigation	(F)	Envelope materials
7.	Shoulder bund	(G)	Drainage
8.	Penman	(H)	Irrigation water management
9.	Quality of irrigation water	(I)	Energy balance
10.	Parshall	(J)	Electrical conductivity

Answers : 1-A, 2-C, 3-F, 4-G, 5-E, 6-D, 7-B, 8-I, 9-J, 10-H

S. No.	Group I		Group II
1.	Estimation of aquifer parameter	(A)	Lang column
2.	Mohr's circle diagram	(B)	Soil conservation structures
3.	Solid granite	(C)	Neutron moisture meter
4.	Euler	(D)	Theis equation
5.	Christiansen	(E)	Aquifuge
6.	Stream bank erosion control	(F)	Uniform coefficient
7.	Dupit-Theim's method	(G)	Shearing resistance of soil
8.	Water harvesting	(H)	Drainage
9.	Moisture measurement	(I)	Submersible pump
10.	Tube well	(J)	Rain fed farming

Answers: 1-D, 2-G, 3-E, 4-A, 5-F, 6-B, 7-H, 8-J, 9-C, 10-I

S. No.	Group I		Group II
1.	Erosivity	(A)	Tube well
2.	Erodibility	(B)	Tube well spacing
3.	Application efficiency	(C)	Rainfall
4.	Radial flow	(D)	Soil
5.	Cone of depression	(E)	Irrigation

Answers : 1-C, 2-D, 3-E, 4-A, 5-B

S. No.	Group I		Group II
1.	Laplace equation	(A)	Mohr's circle
2.	Subsoil tillage	(B)	Pump selection
3.	Hydraulic conductivity	(C)	Stream function
4.	Opportunity time	(D)	Critical flow
5.	Shear strength	(E)	Drainage |
6.	Propeller	(F)	Critical height
7.	Efficiency	(G)	Auger hole
8.	Parshall flume	(H)	Chiesel plough
9.	Wind	(I)	Border irrigation
10.	Cypress creak formula	(J)	Diffuser

Answers: 1-E, 2-H, 3-G, 4-C, 5-A, 6-J, 7-B, 8-C, 9-D, 10-F

S. No.	Group I		Group II
1.	Tile drain	(A)	An instrument used to measure water flow velocity of a stream
2.	Drip irrigation	(B)	Herring bone
3.	Ring basin Irrigation	(C)	A systematic record of different strata maintained during wall digging
4.	Current meter	(D)	Twin fluid nozzle
5.	Sprinkler irrigation	(E)	Rain like irrigation
6.	Well log	(F)	Trickle irrigation in overhead
7.	Resistance block	(G)	Flow measurement in an open field
8.	V-notch	(H)	Stability of slopes
9.	Mohr diagram	(I)	Soil moisture
10.	Hein's apoaratus	(J)	J. Drainable porosity

Answers : 1-B, 2-F, 3-D, 4-A, 5-E, 6-C, 7-I, 8-G, 9-H, 10-I

S. No.	Group I		Group II
1.	Seepage	(A)	Components
2.	Single auger hole	(B)	Hydrograph
3.	Water balance	(C)	Grass water way
4.	Free board	(D)	Energy dissipation
5.	Recession constant	(E)	Hydraulic jump
6.	Apron	(F)	Canal
7.	Rempt cum waste weir	(G)	Moisture measurement
8.	Sequent depth	(H)	Tile drain
9.	Mole plough	(I)	Hooghoudt's method
10.	Neutron moisture meter	(J)	Contour bund

Answers: I-F, 2-1, 3-A, 4-C, 5-B, 6-D, 7-J, 8-E, 9-H, 10-G

S. No.	Group I		Group II
1.	Wind erosion	(A)	Suitable for slope less than 3m
2.	Water erosion	(B)	Suitable for slope more than 3 m
3.	Open well	(C)	Soil conservation measure for hilly area
4.	Tube well	(D)	Used for safe disposal of drainage water
5.	Chute spillway	(E)	Takes water from a shallow depth
6.	Drop spillway	(F)	Stream erosion
7.	pF value	(G)	Strip cropping
8.	Bench terrace	(H)	Saltation
9.	Water erosion control	(I)	Soil tension
10.	Grassed waterway	(J)	Well log

Answers: I-H, 2-F, 3-E, 4-J, 5-B, 6-A, 7-1, 8-C, 9-G, 10-D

S. No.	Group I		Group II
1.	Wind erosion	(A)	Phillips
2.	Water erosion	(B)	Water table
3.	Water pumping	(C)	Bench terrace
4.	Infiltration	(D)	Drainage channel
5.	Rain fal	(E)	Shelter belt
6.	Piezometer	(F)	Laminar flow
7.	Tubewell	(G)	Specific energy
8.	Grassed waterway	(H)	Total head
9.	Froud number	(I)	Draw down
10.	Reynold's number	(J)	Rain gauge

Answers: 1-E, 2-C, 3-H, 4-A, 5-J, 6-B, 7-I, 8-D, 9-G, 10-F

S. No.	Group I		Group II
1.	Rainfall intensity vs time.	(A)	Hydrograph
2.	Cumulative discharge vs time in chronological order	(B)	Hyetograph
3.	Accumulated precipitation vs time in chronological order	(C)	Flow duration curve
4.	Stream flow vs. time in chronological order	(D)	Flow mass curve
5.	Stream discharge vs. percent time the flow in equalled or exceeded	(E)	Mass curve of rainfall

Answers: 1-B, 2-D, 3-E, 4-A, 5-C

S. No.	Group I		Group II
1.	CFTRI	(A)	Jodhpur
2.	CIPHET	(B)	Jhansi
3.	NDRI	(C)	Hyderabad
4.	CIAE	(D)	Dehradun
5.	CAZRI	(E)	Karnal
6.	CRİDA	(F)	Mysore
7.	IGFRI	(G)	Ludhiana
8.	CSSRI	(H)	Bhopal
9.	CSWCR&TI	(I)	New Delhi
10.	IARI	(J)	Karnal

Answers: 1-F, 2-G, 3-E, 4-H, 5-A, 6-C, 7-B, 8-J, 9-D, 10-I

CHAPTER 11

MISCELLANEOUS – AGRICULTURAL ENGINEERING

1. **Soil moisture content:**
(A) Decreases with soil moisture tension
(B) Increases with increase in soil moisture tension
(C) Does not change with soil moisture tension
(D) None of these

2. **The actual area irrigated in a year from an outlet is:**
(A) Irrigation period.
(B) Intensity of irrigation
(C) Irrigation frequency.
(D) None of these

3. **Water horse power of centrifugal pump of 10 litres /sec capacity and 30 meter total head will be equal to:**
(A) 3.0 (B) 5.0
(C) 4.0 (D) 6.0

4. **In drip irrigation design, the design criterion is generally based on an emitter flow variation of:**
(A) <20 per cent (B) <15percent
(C) < 10 per cent (D) <05percent

5. **Bulldozers are frequently employed in cutting and pushing the earth when the hauled distance does not exceed:**
(A) 200 meters (B) 50meters
(C) 100 meters (D) 25 meters

6. **From the hydraulic efficiency point of view, the most efficient cross-section of an open channel is:**
(A) Parabolic (B) Trapezoidal
(C) Semi-circular (D) Rectangular

7. **A soil with exchangeable sodium percentage above 15 and EC of the saturation extract greater than 4mm hos/cm is**
(A) Saline alkali soil
(B) Non-saline alkali soil
(C) Saline soil
(D) Sodic soil

8. **The spacing between lateral tiles lines:**
(A) Decreases with increase in hydraulic conductivity
(B) Increases with increase in hydraulic conductivity
(C) Independent of the hydraulic conductivity
(D) Depends on the outlet conditions

9. **The most suitable method for drilling in boulders, slanted and fissured formations is:**
(A) Core drilling
(B) Rotary drilling
(C) Cable tool percussion drilling
(D) None of these

10. **Leaching efficiency per unit of water is highest for:**
(A) Border irrigation
(B) Basin irrigation
(C) Furrow irrigation
(D) Sprinkler irrigation

11. **For a given head and capacity, a pump having larger specific speed will be:**
(A) Smaller and Cheaper
(B) Larger and costly
(C) Smaller and costly
(D) Larger and cheaper

12. The properties of semi-confined aquifer may be expressed in terms of:
(A) Storage coefficient
(B) Transmissibility
(C) Hydraulic resistance.
(D) All of these

13. Leakage factor has the dimensions of:
(A) Length (B) Time
(C) Velocity (D) Resistance

14. Sprinkler irrigation is not recommended in:
(A) Highly undulating area
(B) Soils having high infiltration rate
(C) High wind areas.
(D) All of these

15. Conjuctive use of surface and ground water may be planned to:
(A) Augment canal supplies
(B) Facilitate irrigation with poor quality of water
(C) Combat water logging
(D) All of these

16. If discharges in two circular orifices O_1 and O_2 placed at depth of 200 mm and 800 mm below a constant level of water in a tank are equal, the ratio of diameters of the orifices O_1 and O_2 is:
(A) √2:1 (B) 2:1
(C) 1:2 (D) 1:√2

17. The dimensions of specific storage are:
(A) $M^0L^3T^{-1}$
(B) $M^0L^{-1}T^0$
(C) $M^0L^2T^{-1}$
(D) $M^0L^1T^{-1}$

18. If flow conditions satisfy the Laplace equation, then:
(A) Flow is rotational
(B) Flow does not satisfy continuity equation
(C) Flow is irrotational but does not satisfies continuity equations
(D) Flow is irrotational and satisfies continuity equations

19. Interceptor drain helps to control water logging by:
(A) Lowering the water table
(B) Preventing subsoil water from reaching the area
(C) Allowing vertical drainage
(D) Draining out excess water to the natural drains

20. The live storage requirement for a reservoir is to be determined by:
(A) Topographical survey
(B) Annual demand analysis
(C) Double mass curve analysis
(D) Mass curve analysis

21. The pressure of water in a confined aquifer is:
(A) Above atmospheric pressure
(B) Below atmospheric pressure
(C) At atmospheric pressure
(D) None of these

22. Delta (△) in cm, duty (in hectare/cume(C) and Base period (in days) are related as
(A) △=864 B/D (B) = 864 △/D
(C) B=864 D/△ (D) D=8.64 B/△

23. Lysimeter is used to measure:
(A) Evaporation.
(B) Evapotranspiration
(C) Atmospheric Pressure
(D) Transpiration

24. In deriving 'Bernoulli's equation' one assumes the fluid to be:
(A) Frictionless and incompressible.
(B) A perfect gas
(C) Flowing in the laminar regime
(D) Flowing in a constant area pipe

25. Which one of the following is not used for measurement of velocity of water flow?
(A) Current meter (B) Pitot tube
(C) Surface floats (D) Tensiometer

26. As per BIS recommendation, the shape of lined canal should be:
(A) Semi circular (B) Trapezoidal
(C) Rectangular (D) Parabolic

27. The porosity of clayey soil usually ranges from:
(A) 30 to 35% (B) 35to50%
(C) 40 to 60% (D) 50 to 60%

28. Soil erosion is more in:
(A) Sandy soils (B) Clay loam soils
(C) Silty soils (D) Clay soils

29. The common conservation measure for agricultural lands are:
(A) Bunding and broad based terracing.
(B) Bench terracing
(C) Contour terracing.
(D) A, B and C

30. The traditional practice for soil and water conservation in Rajasthan is:
(A) Farm pond (B) Check dam
(C) Khadin (D) Gully plug

31. A hyetograph is a plot of:
(A) Rainfall intensity versus time
(B) Rainfall versus time
(C) Cumulative rainfall versus time
(D) Discharge versus time

32. Universal Soil Loss Equation (USLE) is named because:
(A) Its validity is more
(B) It is universally adopted
(C) It accounts all parameters affecting the soil loss.
(D) A, B and C

33. The line joining the points of equal elevation is:
(A) Contour line (B) Reference line
(C) Base line (D) Tie line

34. For the most efficient hydraulic section of a rectangular channel the bottom width is:
(A) Equal to the depth
(B) Equal to 150% of depth
(C) Equal to half the depth
(D) Equal to 200% of depth

35. Rational formula is used to compute:
(A) Design discharge for flat land
(B) Design discharge for slopy land
(C) Design rainfall
(D) Design peak run off

36. Class IV type lands are:
(A) Suitable for cultivation
(B) Not suitable for cultivation
(C) Partly suitable for cultivation
(D) None of these

37. Mohr-Coulomb's failure theory depends upon:
(A) Material failure by shear.
(B) Potential failure plain gives ultimate strength of material
(C) Both (A) and (B)
(D) None of these

38. Watershed is a:
(A) Drainage basin contributing the runoff to a single outlet
(B) Hydrological entity
(C) Unit of area extent in which hydrological cycle completes
(D) All of correct

39. Middle third rule is to safeguard against:
(A) Sliding (B) Crushing
(C) Overturning (D) None of these

40. Contour bunds are adopted in:
(A) Low rainfall regions
(B) highly pervious soils
(C) Where conservation of moisture is needed.
(D) A, B and C

41. The minimum wind velocity required to initiate movement of soil particle is known as:
(A) Critical velocity
(B) Threshold velocity
(C) Dynamic threshold velocity
(D) Intrinsic velocity

42. If W is the width of bench terrace, S is the land slope, then for riser slope of 1:1, the vertical interval is given by:
(A) (100-S)WS.
(B) 0.3 [(WS/2) +2]
(C) 2WS/ (200-S)
(D) None of these

43. A unit hydrograph has:
(A) One unit of direct runoff
(B) One unit of peak discharge
(C) One unit of rainfall duration
(D) One unit of time base of direct runoff

44. Table top bench terraces are suitable for areas receiving:
(A) Medium uniformly distributed rainfall with medium permeable deep soils
(B) Heavy rainfall with permeable deep soils
(C) Low rainfall with permeable deep soils
(D) Very heavy rainfall with permeable shallow soils

45. Soil degradation takes place due to:
(A) Accelerated erosion
(B) Geologic erosion
(C) Glacial erosion
(D) A, B and C

46. The term denudation is used to express the erosion in terms of:
(A) Net erosion per unit area
(B) Gross erosion per unit area
(C) Sediment yield per unit area
(D) None of these

47. Wind erosion is common in:
(A) Humid zones
(B) Arid zones
(C) Arid and humid zones
(D) Arid and semiarid zones

48. V. shaped gullies are found in:
(A) Hilly areas (B) Ravine areas
(C) Flat areas (D) A, B and C

49. The science which deals with the measurement of flow is known as:
(A) Hydrometry (B) Technometry
(C) Hypsometry. (D) None of these

50. Large watersheds are those, in which:
(A) Runoff is major flow
(B) Base flow is major flow
(C) Overland flow is major flow
(D) A, B and C

51. When the soil moisture and rainfall are inadequate during the growing season to support healthy crop growth to maturity is called:
(A) Meteorological drought
(B) Agricultural drought
(C) Hydrological drought
(D) Revenue drought

52. The gully control structure releases water partly through the earthen bund is:
(A) Chute spillway
(B) Drop inlet spillway
(C) Drop spillway
(D) Sub spillway

53. Sheet erosion is also called:
(A) Surface flow erosion
(B) Channel flow erosion
(C) Direct flow erosion
(D) None of these

54. The moisture content at which the drying rate of a product changes from a constant rate to a falling rate is:
(A) Equilibrium moisture content
(B) Critical moisture content
(C) Moisture concentration difference
(D) Drying rate

55. Hammer mill and Burr mill are used for the grinding of:
(A) Milk powder (B) oil seed
(C) Fruit powder (D) Grain

56. Screw conveyer is generally used to move grains horizontal but is can also be used up to an angle of:
(A) 15° (B) 30°
(C) 45° (D) 90°

57. LSU dryer was developed in:
(A) India (B) England
(C) Canada (D) USA

58. Solar energy is free from
(A) Flat plate collectors
(B) Problems of transportation
(C) Focusing Collectors
(D) Natural resources

59. The approximate value of solar constant is:
(A) 1.4 KJ/min. (B) 1.4 KW/m^2
(C) 1.4KWh/min. (D) 1.4 KW/min

60. Which one of the following is a secondary fuel?
(A) Wood (B) Natural gas
(C) Diesel (D) Coal

61. The energy derived from natural resources is called:
(A) Secondary energy
(B) Electrical energy
(C) Primary energy
(D) Potential energy

62. A reflector used in solar power plant to reflect sunlight is:
(A) Convex reflector
(B) Plane reflector
(C) Concave reflector
(D) all above

63. A windmill suitable for lifting of water is:
(A) Single blade type
(B) Triple blade type
(C) Double blade type
(D) Multiple blade type

64. Pump used in the windmill may be:
(A) Reciprocating type
(B) Diaphragm type
(C) Rotary type
(D) all above

65. Which of the following types of biogas plants has digester tank and gasholder as combine unit?
(A) Floating dome type
(B) Fixed dome type
(C) Both (A) and (B)
(D) None of these

66. The use of reflector in solar heating devices is to:
(A) Increase efficiency.
(B) Create green house effect
(C) Decrease efficiency
(D) None of these

67. Calorific value of natural gas ranges from:
(A) 10-15 KJ/kg (B) 33-50 KJ/kg
(C) 20-30 KJ/kg (D) 50-75 KJ/kg

68. Solar radiation reaches the earth in the form of:
(A) Beam radiation
(B) Beam and defuse radiation
(C) Defuse radiation
(D) None of these

69. Ridger is used for:
(A) Making furrows
(B) Making Channels
(C) Earthing up operation
(D) All above

70. In a disc plough, the disc angle influences the:
(A) Depth of cut (B) Forward Speed
(C) Field Capacity (D) Width of cut

71. In Single action disc harrow, two gangs are set an angle of:
(A) Less than 90° (B) Greater than 90°
(C) 90° (D) 180°

72. The optimum operating speed of a rotavator ranges between:
(A) ≤ 100 r.p.m. (B) 400-500 r.p.m.
(C) 150-350 r.p.m. (D) ≥500 r.p.m.

73. The field capacity of a sowing machine depends upon:
(A) Rated width of machine
(B) Field performance index
(C) Forward travel speed.
(D) All of the above

74. Seed drill size is given by:
(A) Seed rate per hectare
(B) Length width of drill
(C) Number of furrow openers x distance between two furrow openers
(D) Area covered per unit time

75. The boiling temperature in the radiator is raised to about:
(A) 90° C (B) 110° C
(C) 100° C (D) 120°

76. The temperature at which fuel catches fire is called:
(A) Pour Point (B) Cloud Point
(C) Flash Point (D) Smoke Point

77. The average threshing capacity of a bullock drawn Olpad thresher per day is:
(A) 150 kg (B) 500 kg
(C) 350 kg (D) 600kg

78. The work of shoe in the reaper is:
(A) to regulate the black
(B) to regulate the width of cut
(C) to regulate the height of cut
(D) to regulate the engine speed

79. Dew point temperature is the temperature at which relative humidity comes to:
(A) 70 Percent (B) 80 Percent
(C) 90 Percent (D) 100 Percent

80. If the moisture content is low, the air is:
(A) cool (B) hot
(C) wet (D) dry

81. The destruction of all micro-organisms in food by thermal processing is known as:
(A) Sterlization (B) Balanching
(C) Pasteurization (D) Scalding

82. Pasteurization of milk is to be carried out at a minimum temperature of:
(A) 125°C (B) 90°C
(C) 71°C (D) 62°C

83. The process of heating and immediately cooling the milk for controlling bacterial growth is known as:
(A) Blending (B) Pasteurization
(C) Regeneration (D) Evaporation

84. Deep bed drying is characterized by:
(A) Ambient temperature drying.
(B) Uniform drying rate
(C) Underground drying structure
(D) Significant moisture and temperature gradients in the bed

85. Electrical resistance blocks made of nylon are most sensitive at soil moisture tensions than:
(A) 15 atmospheres
(B) 10 atmospheres
(C) 5 atmospheres
(D) 2 atmospheres

86. Electrical conductivity of a soil solution is a measure of:
(A) Soil dryness
(B) Soil iron content
(C) Soil salinity
(D) Clay content of the soil

Answers Key

Q. Ans	Q. Ans	Q. Ans	Q. Ans	Q. Ans	Q. Ans	Q. Ans	Q. Ans	Q. Ans
1. A	2. B	3. B	4. A	5. D	6. C	7. A	8. B	9. C
10. D	11. A	12. D	13. A	14. C	15. D	16. A	17. B	18. D
19. B	20. C	21. A	22. A	23. B	24. A	25. D	26. B	27. C
28. A	29. A	30. C	31. A	32. C	33. A	34. D	35. D	36. C
37. A	38. D	39. C	40. D	41. B	42. D	43. A	44. C	45. A
46. C	47. D	48. A	49. A	50. A	51. B	52. C	53. A	54. B
55. D	56. A	57. D	58. C	59. A	60. D	61. C	62. C	63. D
64. D	65. A	66. A	67. B	68. B	69. D	70. D	71. B	72. C
73. D	74. C	75. B	76. C	77. D	78. C	79. D	80. D	81. A
82. C	83. B	84. D	85. D	86. C				

CHAPTER 12

MISCELLANEOUS (SOIL CONSERVATION AND WATER MANAGEMENT)

1. **Evapo-transpiration is measured in the field by**
   (A) Lysimeters
   (B) Ring infiltrometers
   (C) USWB Class-A pans
   (D) Anemometers

2. **Stage recorders measures**
   (A) Velocity of flow
   (B) Height of the water level in channel
   (C) Discharge rate
   (D) None of these

3. **Which of the following is the most accurate instrument for measuring stream velocity?**
   (A) Twin float
   (B) Surface float
   (C) Partially submerged rod
   (D) Current meter

4. **The rational method used for computation of the runoff from watershed gives the**
   (A) Runoff volume
   (B) Discharge rate of channel
   (C) Runoff rate
   (D) Peak runoff rate

5. **The second phase of soil erosion is**
   (A) Rill erosion (B) Sheet erosion
   (C) Gully erosion (D) Wind erosion

6. **Streams which always carry water are called**
   (A) Perennial streams
   (B) Intermittent streams
   (C) Ephemeral streams
   (D) Flow hydrographs

7. **The permissible soil loss under normal conditions lies in the range**
   (A) 30-40 t/ha/yr
   (B) 5-10 t/ha/yr
   (C) 18-35 t/ha/yr
   (D) 50-60 t/ha/yr

8. **Wind erosion is caused by**
   (A) Storms of high intensity
   (B) Mismanagement of land resources
   (C) Clay particles in soil
   (D) Alkalinity of soil surface

9. **The USLE is primarily recommended for calculation of soil erosion from**
   (A) Small watersheds on per storm basis
   (B) Small watersheds on yearly basis
   (C) Large watersheds on yearly basis
   (D) Sediment flow in rivers

10. **Which soil type has a low detachability rate**
    (A) Fine clay (B) Coarse sand
    (C) Silt loam (D) Organic matter

11. **The value of soil erodibility factor in the presence of high organic matter in a soil is**
    (A) Higher
    (B) Moderately higher
    (C) Lower
    (D) No consequence

12. **Land capability class III land can be farmed with conservation measure that are**
    (A) Intensive (B) Moderate
    (C) Below moderate (D) Insignificant

13. Contour faming is recommended for lands with the slope range of
(A) 0-1% (B) 2-7%
(C) 7-12% (D) 12-24%

14. Strip cropping has alternate rows of
(A) Trees and grasses
(B) Trees and row crops
(C) Row crops and close growing crops
(D) None of these

15. Water harvesting, primarily means
(A) Collection & preservation of rainwater
(B) Harvesting water from drains
(C) Diverting water to rivers
(D) Removal of excess water

16. Water harvesting implies storing water in
(A) Reservoirs (B) Rivers
(C) Soil (D) Both A & C

17. A saline soil has
(A) pH > 8.5
(B) pH < 8.5
(C) pH in the range of 8.5 to 10.5
(D) None of these

18. SAR is mainly used for analysis of
(A) Irrigation water
(B) Soil quality
(C) Organic matter
(D) None of these

19. Soil Conservation structures are designed to perform following function
(A) Erosion control
(B) Sediment control
(C) Water conservation
(D) All of these

20. Universal soil loss equation was given by
(A) Dokuchaev
(B) Wischmeier & Smith
(C) Williams
(D) None of these

21. Soil is composite of following compounds
(A) Mineral matter (B) Organic matter
(C) Air and water (D) All of these

22. The process of accumulation of soluble salts is called.
(A) Desalinization (B) Degradation
(C) Salinization (D) All of these

23. The disintegration of rocks and minerals is called
(A) Mechanical weathering
(B) Physical weathering
(C) Chemical weathering
(D) Both A & B

24. The process of conservation of soil NO3 in Nitrous oxide by Pseudomonas
(A) Denitrification
(B) Nitrification
(C) Decomposition
(D) None of the above

25. Role of organic matter in soil
(A) Increase in water holding capacity
(B) Good physical condition
(C) Decrease erosion
(D) All of these

26. Black cotton soil comes in the order of
(A) Entisols (B) Alfisols
(C) Vertisols (D) None of above

27. Passive factors which affect soil formation
(A) Parent material (B) Rainfall
(C) Vegetation (D) Temperature

28. Saline soils are corrected by
(A) Leaching
(B) Calcium carbonate
(C) Gypsum
(D) None of these

29. Which of soil acidity is more harmful?
(A) Reserve acidity
(B) Exchange acidity
(C) Active acidity
(D) None of these

30. Best method of fertilizer application under dry land condition
(A) Broad casting (B) Topdressing
(C) Placement (D) None of these

31. Iron pyrite is generally use to reclaim
(A) Acid soil
(B) Waterlogged soil
(C) Sodic soil
(D) Acid sulphate soil

32. Which Land use class is not suitable for agriculture
(A) V (B) II
(C) III (D) I

33. Class-IV land is represented by the colour
(A) Pink (B) Blue
(C) Red (D) Yellow

34. The watershed programme is normally undertaking in
(A) Irrigated area
(B) High eroded area
(C) Non-mechanized area
(D) Rainfed area

35. CSWCRTI is located at
(A) Jodhpur (B) Hyderabad
(C) Bhopal (D) Dehradun

36. More ravenous area in U.P. are found in
(A) Lucknow division
(B) Jhansi division
(C) Agra division
(D) Meerut division

37. Name the CSWCRTI centre which deals with control measures of Jamuna ravines
(A) Chandigarh (B) Agra
(C) Vasad (D) Mathura

38. A network of gully erosion is called as
(A) Gullied area
(B) Waterlogged area
(C) Hilly area
(D) All of these

39. Amongst all soil constituents organic matter is
(A) Less erodible (B) More erodible
(C) Non erodible (D) None of these

40. Sand dunes are associated with
(A) Wind erosion (B) Water erosion
(C) Chemical erosion (D) Landslides

41. Greater fertility loss is found in
(A) Wind erosion
(B) Water erosion
(C) Chemical erosion
(D) None of these

42. Shelter belt is constructed by
(A) One row of trees
(B) Two rows of tress
(C) More than two row of tress
(D) None of these

43. Common conservation methods for agricultural land are
(A) Bunding &broad based terracing
(B) Contouring
(C) Sowing across the slope
(D) All of these

44. Watershed is a
(A) Drainage basin controlling runoff to a single outlet
(B) Hydrological unit
(C) Both A & B
(D) None of these

45. Relative percentage of sand, silt and clay is known as
(A) Soil texture
(B) Soil structure
(C) Parent material
(D) None of these

46. Application of organic matter increases the
(A) Permeability of soil
(B) Microbial activity
(C) Water holding capacity
(D) All of these

47. Maize crop is an
(A) Erosion permitting crop
(B) Erosion resisting crop
(C) Both A & B
(D) None of these

48. Diversion ditches are constructed for
(A) Diversion of runoff
(B) Removal of flood
(C) Interception & diversion of runoff
(D) None of these

49. Which of the following trees, once very popular in social forestry, is now taken to be environmental hazard?

(A) *Acacia* spp.
(B) *Azadirachta* spp.
(C) *Teminalia* spp.
(D) *Eucalyptus* spp.

50. Species suitable for plywood

(A) Acacia species
(B) Populus deltoides
(C) Tectona grandis
(D) Dalbergia sissoo

51. Basic principles of Soil Conservation is

(A) Protection of soil from impact of rained drop
(B) Increase in wind velocity
(C) Decrease in size of the soil particles
(D) None of these

52. Which of these factors affects the rate of weathering?

(A) Chemical composition of the exposed rocks
(B) Climate
(C) Surface area of exposed rocks
(D) All of these

53. Runoff reduced by

(A) Shallow tillage (B) Deep tillage
(C) Zero tillage (D) None of these

54. The spacing of check dams is kept between

(A) 45-75 m (B) 150-450 m
(C) 80-170 m (D) 75-100 m

55. Soil water is easily available to the plants are called

(A) Gravitational water
(B) Bional water
(C) Capillary water
(D) Hygroscopic water

Answers Key

Q. Ans	Q. Ans	Q. Ans	Q. Ans	Q. Ans	Q. Ans	Q. Ans	Q. Ans	Q. Ans
1. A	2. B	3. D	4. D	5. B	6. A	7. B	8. A	9. B
10. A	11. C	12. *	13. B	14. C	15. A	16. D	17. B	18. A
19. D	20. B	21. D	22. C	23. D	24. A	25. D	26. C	27. A
28. A	29. C	30. C	31. C	32. A	33. B	34. B	35. D	36. C
37. B	38. A	39. A	40. A	41. B	42. C	43. D	44. A	45. A
46. D	47. A	48. C	49. D	50. B	51. A	52. A	53. B	54. *
55. C								

CHAPTER-1

RAINFED AGRICULTURE IN INDIA AND PRODUCTION CONSTRAINTS

The Indian economy is mainly dependent on agriculture, which contributes 21 % of country's capital GDP and 60 percent of employment potential. India made rapid strides in food production during last three decades culminating in self-sufficiency and surplus production. However, feeding the ever-increasing population through the next millennium remains an uphill task. The country will have to feed about 1.3 billion people by the year 2020 requiring 5-6 mt of additional feed grains every year. Besides, the problems of poverty and malnutrition have their own implication to national food security. Over 70 percent of Indian population, which is predominantly rural, do not have proper access to food and non-food commodities due to poor employment and infrastructure facilities. This reminds all those concerned with the country's food security of the daunting task ahead in order to ensure access to food to the growing population. Rainfed agriculture occupies 67 percent of net sown area, contributing 44 percent of food grain production and supporting 40 percent of the population.

Constraints of production in rainfed areas

The rainfed lands suffer from a number of biophysical and socio-economic constraints which affect productivity of crops and livestock. These include low and erratic rainfall, land degradation and poor productivity, low level of input use and technology adoption, low draft power availability, inadequate fodder availability low productive livestock and resource poor farmers and inadequate credit availability.

Strategies for Sustained Food Production in Rainfed Region Identification of viable rainfed technologies

A number of economically viable rainfed technologies have been developed over the years in the country to address the problems of food production in rainfed agriculture through CRIDA and its network centres for the last three decades. These technologies have been evolved after refining them in farmers' field through Operational Research Projects, Institute Village Linkage Program (IVLP) and farm science centres. These include simple practices like off-season tillage in rainfed Alfisols and related soils for better moisture conservation and weed control. Farmers in Operational Research Project (ORP) areas of Hyderabad adopted this practice in sorghum and castor and realized yield advantage by 40 percent over traditional practice. Lack of adequate draft power with many small farmers, however, is one of the major constraints to popularize this practice. Custom hiring of tractor is effective solution of farm mechanization on these lands.

Soil and rain water conservation techniques

Efficient conservation of rainwater is the central issue in successful dryland farming. Extensive trials conducted by the soil conservation and dryland research centres have led to the identification of a number of inter-terrace land treatments besides contour and graded bunds. These techniques are location specific and the benefits from their adoption are highly variable depending on the rainfall intensity, slope and texture of the soil besides the prevailing crop/cropping system. Farmers have not widely adopted mechanical measures like contour bunds, graded bunds, grassing of waterways and construction of farm ponds without the government support due to financial constraints. However, studies at Hyderabad, Bangalore and Anantapur revealed that more than 80 percent farmers follow simple conservation measures like sowing across the slope, opening of dead furrows and key line cultivation. The yield improvement by adoption of soil and water conservation measures vary between 12 and 20 percent which are at times not convincing enough to farmers. However, cumulative effects are significantly visible at some locations. Since such measures help in long-term conservation of resources, these are implemented through the Government of India or the respective State Government sponsored watershed management programmes.

Timely planting of crops

Timely sowing and precision are essential for getting good plant stand, higher yield and optimum utilization of rainfall and reduction in the incidence of pests and diseases. A number of demonstrations have been taken up in farmers fields through ORPs, KVKs and IVLP programmes in different rainfed regions of the country .In case of sorghum and castor in farmers fields of Hyderabad, a fifteen day delay in sowing led to reduction of 300 and 850 kg/ha compared to normal sowing. Inadequate availability of farm implements and draft are major constraints. However, seeding and intercultural experiments developed by CRIDA and AICRPDA centres helped in overcoming the constraints to some extent.

Adoption of improved crop varieties

A number of improved varieties and hybrids were evaluated in the farmers fields to identify suitable ones for matching growing periods for inter and sequence rainfed cropping systems. For example, farmers gained additional benefit ranging from Rs. 2000- 4000/ha by adopting improved varieties of sorghum, castor and sunflower in Alfisols of Hyderabad.

Efficient crops and cropping systems

To achieve appropriate land use, efficient inters and sequence-cropping systems were recommended based on soil type, rainfall and length of growing seasons. The studies at Hyderabad indicated only 25 percent farmers adopted 2:1 ratio of sorghum-pigeonpea. Whereas 45 percent of farmers adopted the finger millet + pigeonpea system (8:1) ratio in Alfisols of Karnataka and maize + soybean system (2:2) was accepted by Ranchi farmers. Groundnut + pigeonpea (7:1) was widely accepted by the farmers in Rayalseema of Andhra Pradesh. Some of the constraints for wider adoption by the farming communities are preference for fodder genotypes in cereals rather than grains for feed to live stock, lack of suitable farm implements to seed in different ratios, delay in planting of kharif for double cropping systems. These have to be refined under on-farm situations for greater acceptance by the farmers.

Farm implements

Proper tillage and precise placement of seed and fertilizers in the moist zone are most critical to for successful crop establishment in drylands. Since the sowing of crops must be completed in a short span of time, use of appropriate implements is necessary to cover large area before the seed zone dries out. Suitable implements have been recommended for various locations to meet this requirement. These are designed to suit the soil type, crop and the draught power availability. In many cases, the existing local implement used by the farmers have been improved to increase their working efficiency.

Nutrient management

Fertilizer recommendations in rainfed crop production have been made primarily for NPK along with the conjunctive use of chemical, organic and bio-fertilizer. Inclusion of legumes in cropping systems can supplement fertilizer N to the extent of about 20 kg N per ha.

Integrated pest management (IPM)

Pests and deceases constitutes a major constraint to increased food production. Crop losses due to pest attack range from 10-30 percent depending on the crop and environment. Complete crop failure may occur in case of serious attack. Indiscriminate use of the pesticides in rainfed crops will lead to harmful side effects such as direct toxically to the applicator or consumer, development of strains or pests resistant to pesticides, resurgence of pest species, outbreak of secondary pesticides, and destruction of non-target organisms such as parasites and predators and accumulation of harmful residues of food products. Integrated pest management is one of the alternatives for the chemicals used for pest management. IPM encourages the most comfortable and ecologically sound combination of available pest suppression techniques and to keep the pest population below economic threshold. Easily adaptable and economically viable integrated pest management strategies have been developed for the control of major pest in rainfed crops like cotton and pulses.

Alternate Land use Systems

Despite evolving a number of production technologies, arable cropping in drylands continues to suffer from instability due to aberrant weather. To provide stability to farm income and also utilize the marginal lands for production of fodder, fuel wood and fibre, a number of alternative land use systems were evolved based on location specific experimentation.

Integration of livestock with rainfed farming systems

Live stock is treated as a part of farming system in rainfed agriculture in India. The soil, plant, animal cycle is the basis for all feed used by the animals. The livestock in the rainfed regions are weak. Farmers in this area often sell their cattle due to the scarcity of fodder. In India the land holdings are being reduced with increased population pressure. Hence, land not suitable for agriculture has to be diverted for raising fodder need of animals through the appropriate alternate land use system such as improved pasture, silvipasture, hortipasture and tree techniques.

Integration of the technologies through watershed approaches

The concept of watershed is important in efficient management of water resources. As the entire process of agricultural development depends upon the status of water resources, the watershed with distinct hydrological boundary is considered ideal for taking up a development programme. In brief, planning and designing of all soil conservation structures are carried out considering the peak runoff. In this context, the watershed concept is of practical significance. Also, the entire development needs are to be taken up on topographic considerations from ridge to valley.

Resource Conservation Measures

Details about conservation measures adopted in cultivated lands have been delineated. Based on the nature and type of barriers and their cost, the conservation measures in arable lands can be divided into three categories: (i) Hardware treatments (ii) Medium software treatments and (iii) Software treatments.

Farming system approach

Of late, it has been increasingly recognized that unlike irrigated areas, it is difficult to develop profitable technologies for heterogeneous agro-ecological and socio-economic conditions of small holders in arid and semi-arid regions. Since, the problems are complex ,addressing only a component of the farming system, e.g. crop variety, fertilizer use or even crop husbandry per se is not expected to bring about a significant increase in the productivity as witnessed in irrigated areas. The extension strategy should be such as to match this challenge. The farming systems perspective, dovetailed on watershed approach therefore can be the appropriate management strategy for such regions. The following steps constitute the farming systems mode for research, both on-station and on-farm (Watershed).

- PRA and assessment of socio-economic conditions of people.
- Identification of ITK (indigenous technical knowledge).
- Collection of available technological knowledge on various components of the farming system – arable farming, animal husbandry, water harvesting, management of wastelands and alternate land use systems etc.
- Focus group (farmers) interaction to identify appropriate technology for different categories of farmers.
- Identification of lead farmers to function as facilitator in technology application and adoption.
- Identification of points of synergy among systems components.
- Structuring of technological components with maximum synergy.
- Phasing of program over the project period.

CHAPTER 2

SOIL AND WATER CONSERVATION IN RAINFED AREAS

Globally 69% of all cereal area is rainfed, including 40% of rice, 66% of wheat, 82% of maize and 86 % of other coarse grains. Worldwide, rainfed cereal yield is 2.2 metric tons per ha, which is about 65% of the irrigated cereal yield (3.5 metric tons/ha). Rainfed areas currently account for about 58% of world cereal production. In Asia, about 700 million ha area is rainfed, and nearly 70% of the population depends on agriculture for a living. The current productivity of Asian rainfed agriculture is around 600 to 1000 kg per ha depending on the eco-region and crop (ICRISAT Report). Poverty incidence is 28%. In India, nearly 67% of the net sown area is rainfed. Rainfed Agriculture is crucial to the country's economy and food security because about 44% of the total food production is contributed by (67%) rainfed agriculture.

The annual long-term average rainfall of India is 117 cm compared to the global average of 110 cm. It is highly erratic in its spatial and temporal distribution. Of the total precipitation that falls in the country, less than 29% is harvested or utilized and 71% of it is lost to the sea. It is also the hill eco-system which is associated with higher rainfall. This is the reason why soil erosion rates are very high in this eco-system. For example, in the state of Himachal Pradesh, more than 30% i.e. almost one third of the geographical area, has soil erosion rates more than 10 t/ha/year. This not only results in the loss of soil but also nutrients. It is estimated that at the national level, 5.3 billion tonnes of soil and 6 million tonnes of nutrients are lost resulting in an economic loss in production of more than Rs. 11,000 crores.

Characteristics of Rainfed Areas

Ecologically, rainfed areas are diverse, ranging from deserts in plains to humid rainfed areas in hill and mountain ecosystems. Traditional farming is practiced and the crops mature at different times. Agriculture and animal husbandry are not practiced in isolation but in combination as agri-silvicultural, agri-pastoral and agri-silvi-pastoral systems. The cultivars may be individually poor in productivity but collectively the crop mixes are reasonably productive. Rainfed agriculture is generally characterized by low productivity, the decrease of the agricultural intensification and the instability in production and the farmers' income.

Rainfed agriculture is constrained by harsh climates, erratic rainfall patterns, low soil productivity and poor socio-economic conditions of the farmers. The degree of limitations, constraints and potentials of rainfed areas vary over a wide range with the landscape characteristics, climatic factors (precipitation, and micro-climate) and socio-economic factors (ethnicity, type of economic activity, resource management pattern etc.). The problems associated with rainfed agriculture are more acute in degraded lands, which form a sizeable part of rainfed areas; almost fifty percent of the land in Asia today is classified as degraded. Judicious water management and sustained use of soil resources are the key elements to improve rainfed agriculture.

Why Research in Rainfed Areas?

There are four strategic reasons why rainfed agricultural research should gain attention:

1. Since the prospects for continued expansion of output from the irrigated areas are limited, the logical option for improved food security is to turn to rainfed agriculture.
2. The land use pattern is changing. Rapid industrialization and urbanization are impinging on fertile agriculture lands, and mainly on irrigated areas. The answer to improved agricultural output and food security is again to focus on rainfed agriculture.
3. Rainfed areas support very fragile and extremely useful ecosystems. With the growing demand for agricultural land, the pressure on rainfed areas has begun to increase. It has become essential that these fragile areas be managed in a sustainable manner.
4. Rainfed areas support the poorest of the poor and are home to many indigenous ethnic minorities who themselves are generally poor. The focus on rainfed areas is an important strategy to support resource poor communities and help alleviate poverty.

Soil and Water Conservation in Rainfed Areas

Soil and water conservation are in fact simultaneous processes. Conservation of one in general leads automatically to the conservation of the other resource.

1. Soil Conservation

Soils in hills must be protected against soil erosion by adopting mechanical, vegetative and/or improved agronomic practices. There are several site-specific options of soil conservation available.

The various remedies to combat erosion on grassland and hilly terrain are:

Riparian fencing: by fencing waterways, cattle won't trample river banks and natural vegetation can re-grow. Also tree planting on river sides helps. Make sure the flat land is not part of the river's flood zone.

Shelter belts: shelter belts provide roots at the boundaries of the field, sheltering the soil, supplying organic matter.

Grassed waterways: a very effective protection of shallow water ways is to grass them over. Grass leaves bend in the current, protecting the soil. Note that grass needs sunlight, so shelter belts must be placed on the correct side. Make grassed waterways wide and shallow to slow the water down and to prevent it from rilling and then gullying the waterway.

Spaced tree planting: trees provide deep roots that hold the soil; cycle deep nutrients and lost nutrients, provide organic matter and more. Trees must compete with highly metabolizing grassland, thus should be fast growing and deciduous.

Debris dams: debris dams check the water in steep gullies. They should not be used in slow-flowing waterways. Any kind of debris would do but natural products such as tree stumps, branches, plantings are better.

Retirement fencing: problem areas should be fenced for retirement so that seedlings are no longer grazed. Fencing is an absolute must to encourage natural growth and to protect tree plantings.

Close tree planting: on retired land, trees can be planted closely. Once retired, any kind of tree can be planted, as necessary for forestry or for slow growing native bush.

Top-dressing: fertilizing is the best remedy against erosion, both immediately and in the long term.

Remedies for saving croplands are:

Shelter belts: reduce wind and evaporation, increase soil temperature.

Reduced tillage: tilling only the areas that matter while minimally disturbing the soil. Tilling between furrows.

No tillage: the ultimate in reduced tillage. Weeds are controlled by cutting or with herbicides. The natural soil organisms are allowed to rebuild the soil, increasing its porosity and resistance to compaction, while better retaining nutrients. Only where seeds are planted is the soil 'drilled'.

Stubble mulching: leaving stubble on the field as long as possible to reduce evaporation, to catch snow in winter, to keep the soil covered. Rather than ploughing the stubble under, which would disturb the soil while inviting soil organisms to decompose the organic matter far too rapidly, the stubble is mulched to leave soil organisms the task of digging it under. This reduces weed growth and keeps the soil covered, while feeding the soil biota slowly.

Contour ploughing: works a bit like terracing, preventing moisture from running down-hill and reducing erosion considerably.

Strip cropping: reduces the down slope length of the field considerably, reducing the chance of riling. Increases biodiversity. Leached nutrients are used by neighboring strips. Shelters bare strips from wind and wind erosion. Makes better use of the land but is more labour intensive.

Terracing: the ultimate in soil conservation on steeper hills. Terracing can reduce erosion several-fold.

Reduced compaction: using machinery and technology that spreads its weight over a larger area. Reduced tilling causes soils that resist compaction better.

Optimal fertilizing: although fertilizer is an important ingredient for increasing profit and yield, it can also degrade the land in many ways. Regular soil tests are one way of overcoming this.

Water harvesting: by keeping forests planted above the field, one prevents runoff from reaching the cropland. Trees also harvest water in their very porous soils. Water is slowly released into the water table, reaching the cropland down-slope. Ponds can be made by damming the upper catchment area of a creek. Water will flow through the water table to water cropland down-slope. Water can be used for irrigation and stock drinking water.

2. Soil-moisture conservation

Rice, maize and wheat are three cereal crops cultivated by hill farmers as their staple food. Water scarcity at the time of sowing of crops is a major problem. During *kharif* (wet) season, delayed onset of monsoons delays the sowing of crops, resulting in poor crop establishment and lower productivity. Maize is very sensitive to sowing date. The maturity of late transplanted rice in hills invariably coincides with low temperatures, resulting in increased spikelet sterility; delayed planting of rice further accentuates the problem. Fallow period of about 30-45 days between the harvest of maize/rice and sowing of wheat leaves residual soil moisture insufficient for optimum germination of wheat seeds. Early withdrawals of monsoons further accentuate the problem. Crop yields in rainfed areas, therefore, are very low and highly unstable; crop failures are a common feature in rainfed areas. Under these circumstances, *in situ* conservation of rainwater helps in the timely sowing of crops. There are several options of *in-situ* soil moisture conservation.

2.1 Land shaping

On gentle sloping lands minor leveling of individual fields with minimum cut and fill is an effective method of soil and water conservation. Each field should be provided with a suitable outlet for drainage of excess water. On level lands the individual fields should be provided with suitable bunds to retain the rain water *in situ.*

2.2 Mulching

Mulching for soil moisture conservation is an age-old practice and is selectively used by the farmers in hills. Several materials are available which can be used as mulch, such as organic or inorganic materials of natural (field crop residues, waste plant biomass, wood chips, saw dust, coco-coir etc.) or synthetic origin (plastic sheets, soil conditioners), soil (dust mulch) or pebbles (stone mulch). The choice of a mulch material, however, depends on its cost, availability, ease of handling and efficacy etc.

Technology was perfected at Himachal Pradesh Agricultural University, Palampur, for *in situ* soil moisture conservation by using the concept that mulch should be applied when the soil profile is almost saturated with water and the evaporation is controlled by atmospheric evaporativity. We used waste plant materials (*Lantana, Eupatorium, Ipomea, Shorea robusta* etc.) as mulch to during rainy (*kharif*) season to enhance residual soil moisture for timely sowing of winter (*rabi*) crops. Maize stalk can also be used as mulch at maize harvest until sowing of wheat for soil moisture conservation. The maize stalk is removed from soil surface after sowing of wheat and used as animal feed.

2.3 Organic matter build-up in soils

Soil organic matter plays three important functions: nutritional function (supplies essential nutrients), biological function (food for microbes), and soil physical, chemical and physico-chemical function (improvement in soil properties). Water holding capacity (WHC) and available water capacity (AWC) of soils are directly linked to SOC. Studies have shown that 1% increase in soil OM increases field capacity by about 2.2%, permanent wilting point by about 1%, and available water capacity by about 1.5%. Soil incorporation of organic materials such as FYM, compost, green manures, waste plant residues, farm weeds or even crop residues improves SOC and soil water retention.

2.4 Conservation tillage

Conventional tillage (repeated primary and secondary tillage operations) invariably enhances OM decomposition by breaking soil aggregates and exposing the soil to weathering agents, thus, weakening soil structure, and making soil loose to be washed away with runoff water. Conservation tillage (a non-inversion tillage that retains at least 30% of the soil surface covered by residues after planting, and is less intensive than conventional tillage) has been found as a potential alternative to conventional tillage to reduce soil erosion by water and improve and sustain soil health. The term "conservation tillage" is an umbrella classification for several crop production systems, including no-till, reduced-till, minimum-till, direct seeding and other names. If tillage operations are inevitable, they should be performed across the slope so as to retain water at soil surface and cut down the flow velocity of runoff water.

2.5 Alley cropping

The alley cropping is another technique suited particularly to hills for soil and moisture conservation. It is essentially an agro-forestry system in which food crops are grown in alleys formed by hedgerows of trees and shrubs. The hedgerows are cut back at planting and kept pruned during cropping to prevent shading and to reduce competition with food crops. When there are no crops, the hedgerows are allowed to grow freely to cover the land. It can easily be adopted by resource-poor farmers in hills and is ideally suited to marginal areas. Tree and shrubs in the alley system

- ✓ Provide green manure or mulch for companion crops. In this way plant nutrients are recycled from deeper soil layers.
- ✓ Provide prunings, applied as mulch, and shade during the fallow to suppress weeds.
- ✓ Provide favourable conditions for soil macro- and micro-organisms
- ✓ When planted along the contours of sloping lands, provide a barrier to control soil erosion.
- ✓ Provide prunings for browse, staking material and fire wood.
- ✓ Provide biologically fixed N to the companion crop.

The major advantage of alley cropping over the traditional shifting cultivation and bush fallow systems in hills is that the cropping and fallow phases can take place concurrently on the same land, thus allowing the farmer to crop for an extended period without returning the land to bush fallow.

3. Rainwater harvesting

An old technology is gaining popularity in a new way. Rain water harvesting is enjoying a renaissance of sorts in the world, but it traces its history to biblical times. Extensive rain water harvesting apparatus existed 4000 years ago in the Palestine and Greece. In ancient Rome, residences were built with individual cisterns and paved courtyards to capture rain water to augment water from city's aqueducts. As early as the third millennium BC, farming communities in Baluchistan and Kutch impounded rain water and used it for irrigation dams.

Rainwater harvesting is the process of collecting, concentrating and improving the productive use of rainwater, and reducing unproductive depletion. There are two main techniques of rain water harvestings:

- Storage of rainwater on surface
- Recharge to ground water

Surface storage of water may be achieved either through *in-situ* rainwater harvesting or by collecting rainwater from a catchment area and channeling the runoff and using it to increase the available water in a relatively smaller growing area. Traditionally, the structures used for storage of rain water on surface are underground tanks, ponds, check dams, weirs etc. The source of water may be direct rain water, runoff from rooftops, springs, small streams, and/or seasonal floodwaters from local streams etc. Even water flowing in *kuhls* may be temporarily stored in small tanks constructed along the course of the *kuhl,* called the auxiliary tanks, and the water is used during lean periods or critical crop growth stages. The rainwater may be harvested and stored in large reservoirs on community basis or in small (100-200 m^3 capacity) farm ponds constructed on individual farmers' fields. The choice depends basically on the objective, agro-ecological situation, and people willingness and participation.

3.1 *In-situ* rainwater harvesting

There are several options:

a. **Land leveling and field bunding:** Level the individual plots, and construct thick, strong, raised bunds to impound rainwater *in situ*. Improve infiltration and water holding capacity of soils through proper soil management (use organic materials and proper tillage), and water uptake by plants.

b. **Land shaping in orchards:** Adopt land shaping in orchards (trenching, providing directional slope for flowing water etc.) in such a way that rain water concentrates into tree basins.

c. **Micro-catchment water harvesting:** The flat land is converted into ridge-furrow system and crops are planted in furrows with the objective of concentrating rainfall in planting area and root zone. Sowing of crops in furrows rather in flat beds helps in harvesting rain water, improves moisture conservation and enhances crop yields.

d. **Raised-sunken bed technology:** Cultivate high water requiring crops like rice in sunken beds and low water requiring crops (maize, soybean, vegetables etc.) on raised beds. A raised : sunken bed ratio of 1.5:1 was found ideal for rice-based cropping system under high rainfall (Palampur) conditions. Beds are semi-permanent in nature and do not require frequent recurring expenditure on their maintenance. Raised beds are also suitable for operations with animal drawn implements. This technology was also found useful for rainfed Vertisols of high rainfall areas.

e. **Terracing:** Adopt terracing on sloping landscapes. On terraced lands, grow crops requiring well drained conditions (maize, soybeans, vegetables etc.) on upper terraces and high water demanding crops (rice, vegetables) on lower terraces where run-off concentrates. Divert water from upper terraces to lower terraces.

3.2 Rainwater harvesting in farm ponds

Rainwater may be harvested in farm ponds and used as supplemental irrigation during lean periods. Water harvesting in relatively small ponds (100-200 m^3) is ideal for hill farmers who have small and fragmented land holdings. In addition, they do not require elaborate management skills and resources. Different types of farm ponds (shape, size, mode of construction) may be constructed at a suitable location in the farm and lined against seepage losses. The ponds may be lined with polyethylene sheet (covered with soil or bricks or round boulders etc. to protect it against UV sun-light), silpauline sheet (UV resistant), bitumen, cement-concrete lining, RCC etc. A line diagram of a small poly-lined farm pond is shown in Fig 1.

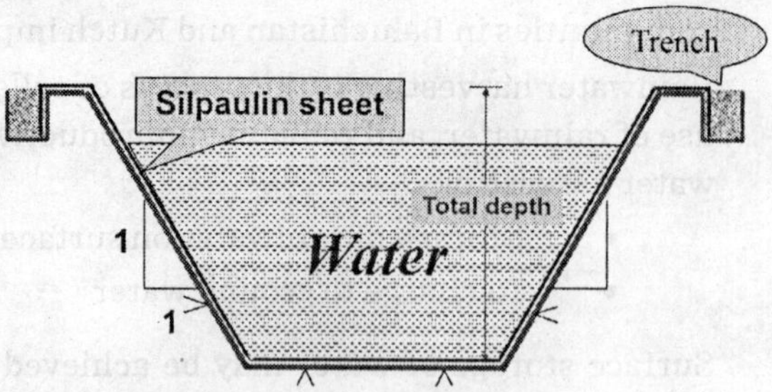

Figure 1.: Shape of a Farm pond with 1:1 side stop

The cost and longevity of structure will depend on the lining material used. Cost of construction of dug-out farm ponds with different lining materials incurred at Palampur is given in Table 1. It may be noted that the cost of lining may vary over a wide range depending on the location and availability of the material etc.

Table 1. Cost of construction of dug-out farm ponds with different lining materials.

S. No.	Lining material	Cost (Rs/m^3 water harvested)
1.	Cement -concrete	709
2.	Silpaulin sheet (UV resistant)	348
3.	Black polyethylene sheet covered with bricks	419
4.	Bitumen lining	575
5.	Black polyethylene sheet covered with soil	277
6.	Mud	157

Capacity of ponds varied between 6.33 and 17.6 m^3

3.3 Watershed approach

Adoption of appropriate soil and water conservation practices on a watershed basis is considered to be the only way to control soil erosion and improve our environment in the mountainous regions. The measures are to be adopted in conformity with the concept of integrated land use planning for development and improvement of catchment and command area. Efforts must be directed towards utilizing the maximum amount of rain to meet the human, animal and crop needs and at reducing to the minimum the damage by floods and soil erosion. Excess water should be stored in the catchment areas which will reduce the fury of flash floods, recharge the ground water and improve the environment. Runoff-collection ponds in the catchments, though they might get silted up in a few years, will be more useful than the measures in the lower reaches. To prevent rapid siltation of tanks, the contributing catchments (even if they are not cultivated but used for grazing or forestry purposes) need to be well managed so that soil erosion is prevented. All common lands should be put under fuel/fodder trees. Planting of barren areas, especially on slopes, with grass cover is an important component of integrated watershed management programme. Grazing should be completely restricted. After the area is completely protected from grazing, better grasses can be planted. The grasses of industrial importance should also be planted so that there is some economic return to the farmers as well. The surface vegetative cover will not only protect the land from the beating action of rain drops and bind the soil particles but would also decrease the velocity of flowing water and cause less of soil erosion.

The watershed approach, however, has to be followed on a community basis and no individual farmer can do it. This program is difficult to implement without the help of government agency and non-government organizations. Also the willingness and participation of local people is a must for the success of any watershed development program. There has to be a proper coordination among state departments (Forest, Agriculture, Soil Conservation, Horticulture and Animal husbandry etc.), NGOs, and local community for the implementation of watershed program on a community basis.

Common principles have been developed by Government of India and circulated among different field agencies at district level. These must be adhered to while implementing the watershed development programmes.

4. **Groundwater Recharging**

Surface water is inadequate to meet our demand and we have to depend on ground water. Increased dependence on groundwater for irrigation and other uses have over-exploited this natural water resource. Even in hills, water table has started showing decline. Further, due to rapid urbanization, infiltration of rain water into the sub-soil has decreased drastically and recharging of ground water has diminished.

Structures generally used for recharging ground water are:

Pits: Recharge pits are constructed for recharging the shallow aquifers. These are constructed 1 to 2 m, wide and to 3 m deep which are back filled with boulders, gravels, coarse sand.

Trenches: These are constructed when the permeable stream is available at shallow depth. Trench may be 0.5 to 1 m wide, 1 to 1.5 m deep and 10 to 20 m long depending up availability of water. These are back filled with filter materials.

Dug wells: Existing dug wells may be utilised as recharge structure and water should pass through filter media before putting into dug well.

Hand pumps: The existing hand pumps may be used for recharging the shallow/deep aquifers, if the availability of water is limited. Water should pass through filter media before diverting it into hand pumps.

Recharge wells: Recharge wells of 100 to 300 mm diameter are generally constructed for recharging the deeper aquifers and water is passed through filter media to avoid choking of recharge wells.

Recharge Shafts: For recharging the shallow aquifer which are located below clayey surface, recharge shafts of 0.5 to 3 m diameter and 10 to 15 m deep are constructed and back filled with boulders, gravels & coarse sand.

Lateral shafts with bore wells: For recharging the upper as well as deeper aquifers lateral shafts of 1.5 to 2 m wide & 10 to 30 m long depending upon availability of water with one or two bore wells are constructed. The lateral shafts are back filled with boulders, gravels & coarse sand.

Spreading techniques: When permeable strata start from top then this technique is used. Spread the water in streams/*Nalas* by making check dams, *nala* bunds, cement plugs, gabion structures or a percolation pond may be constructed.

Construction activity in and around the city is resulting in the drying up of water bodies and reclamation of these tanks for conversion into plots for houses. Free flow of storm run off into these tanks and water bodies must be ensured. The storm run off may be diverted into the nearest tanks or depression, which will create additional recharge.

5. **Rejuvenation of traditional water sources**

Efforts should be made to protect the traditional water resources, such as wells, *baudies*, springs, ponds etc. against drying, pollution or physical destruction. Government is encouraging local communities through *panchayats* etc. for taking interest and responsibility of rejuvenating and maintaining such water sources.

Water springs are important natural sources of water for hill people both for domestic and agricultural purposes. Water springs in hills are drying due to deforestation, mining, blasting activities in the catchment areas (road construction, queries or other infrastructural development etc.), and declining rainfall events. Deforestation increases runoff and decreases infiltration, thereby resulting in relatively less groundwater recharging. Blasting causes drying of water springs and lowering of water table in wells by disrupting water-flow lines. For rejuveniation of natural water resources in hills, deforestation must be curved, land must be put under vegetation and blasting activities reduced.

6. **Efficient utilization of water resources**

Farmers must follow improved agronomic practices to cultivate crops in order to sustain soil fertility, achieve maximum crop productivity and enhance water productivity. Some such practices are:

- Select crops and crop cultivars as per reliable irrigation sources
- Cultivate 'low volume high value crops'
- Adopt scientific irrigation schedules (when, how and how much to irrigate) and methodology for irrigating crops
- Avoid under and over-irrigation
- Use improved irrigation practices (sprinklers and drip irrigation) for maximizing water-use efficiency
- Use fertigation
- Adopt scientific agronomic practices (optimum fertilizer use based on soil test values, weed control, plant protection measures etc.)
- Adopt soil and water conservation measures
- Adopt protective agriculture

CHAPTER 3

SOIL AND LAND RESOURCES

What is Soil?

Soil is a natural body consisting of layers (soil horizons) of primarily mineral constituents of variable thicknesses, which differ from the parent materials in their texture, structure, consistence, colour, chemical, biological and other physical characteristics.

"Soils are applied solely to those superficial or nearly superficial horizons of rocks, that have been more or less modified naturally by the interaction of water, air and various kinds of organisms, either living or dead; this being reflected in a certain manner in composition, structure and colour of such formations. Where these conditions are absent, there are no natural soils, but either artificial mixtures or rocks."

— By Dokuchaev

"Soil is a dynamic natural body developed as a result of pedogenic processes through weathering of rocks, consisting of mineral and organic constituents, possessing definite physical, chemical and biological properties having a variable depth over the surface of the earth and providing a medium for plant growth."

— By Velayutham and Bhattacharya, 2000

"Soil is the more or less loose and crumby part of the outer earth crust in which, by means of their roots, plants may or do find foot hold and nourishment as well as all other conditions essential to their growth."

— By Hilgard

Soil may be defined as "A dynamic natural body on the surface of the earth in which plants grow. Composed of mineral and organic materials and living forms."

— By Buckman and Brady

Soil Components:

Soils consist of four major components i.e. (1) mineral matter (2) organic matter (3) water, and (4) air. The relative proportions of these four soil components vary with soil type and climatic conditions. A review the approximate proportions (by volume) of the four soil components in a mineral soil under optimum conditions for plant growth.

- Soil contains about 50% solid space and 50% pore space.
- Total solid space of soil is occupied by mineral matter and OM by about 45% and 5%, respectively.
- Total pore space of the soil is occupied by air and water by 25 and 25 %, respectively
- OM content is very small and varies from about 3-5% by wt. in a top soil.

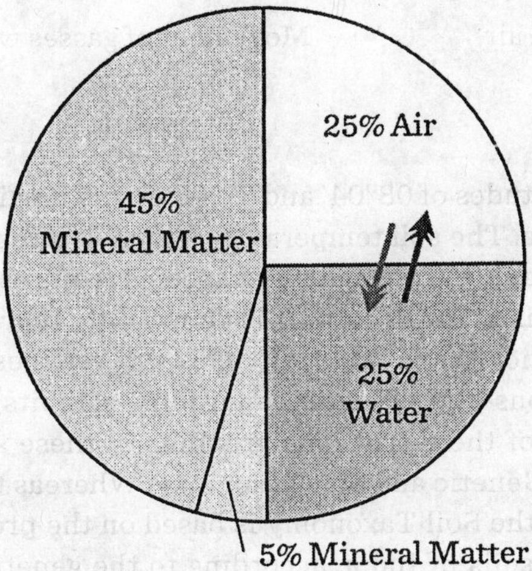

1. **Mineral components**

   Mineral particles are inorganic materials derived from rocks and minerals. They are extremely variable in size and composition.

   a. Composition
      - Types of primary and secondary minerals
      - Weathering
      - Most common elements in soils
      - Particle and bulk density

   b. Size of soil particles
      - Soil texture

   c. Soil colloids
      - Types of soil colloids
      - Specific surface area
      - Charges on soil colloids

2. **Organic components**
   - Non-humic substances
   - Humic substances

3. **Soil water**

   Water is the major transport agent for fluxes within and between terrestrial ecosystems. It is a prerequisite for all active life, and participates in geochemical cycles by weathering geological substrates, by leaching materials to groundwater and by moving ions and particles through the soil profile.
   - Soil water content
   - Soil solution
   - Soil water flow
   - Qualitative description of soil wetness

4. **Soil air**

   Soil pores, the voids between minerals, organic matter, and living organisms, are filled with air or water. There is a dynamic equilibrium between water and air content within a soil. When water enters the soil, it displaces air from some of the pores.

- Composition of soil air
- Movement of gasses within soil
- Soil porosity

SOILS OF INDIA

India, situated between the latitudes of 08°04′ and 37°06′ N and longitudes of 68°07′ to 97°25′ E, has a geographical area of 329 M ha. The soil temperature and moisture regimes play an important role in classifying soils and crop planning, as each plant species requires, specific soil temperature and moisture conditions for its optimal growth. India is dominantly represented by thermic and hyper-thermic temperatures and Ustic, Aridic and Udic moisture regimes. India with a wide variety of landforms, geological formations and climatic conditions, exhibits a large variety of soils. India represents all the major soils of the world. The grouping of these soils can be achieved by using either of the two systems, viz. Genetic and Soil Taxonomy. Whereas the Genetic System is based on genetic factors and processes, the Soil Taxonomy is based on the properties of soils, which are the reflection of soil genesis. The soils of India, according to the genetic factors have been classified into eleven major soil groups.

- Alluvial soil
- Black cotton soil
- Red soil
- Laterite soil
- Forest and mountainous soil
- Organic soil
- Skeletal soil
- Desert soil
- Tarai soil
- Podzole soil
- Saline-sodic soil

Alluvial soils

The name alluvial is given to the soils that have developed in alluvium, irrespective of their place of occurrence and degree of development. They are one of the important groups of soils for agricultural production. They are extensively distributed in the states of Rajasthan, Haryana, Punjab, Uttar Pradesh and in valley of Krishna, Kaveri, Brahmputra, Godawari, Narmada, Tapti and Surma rivers.

Order	:	New alluvial soil – Entisols order Old alluvial soil – Inceptisols order
Area	:	75 M ha (1st largest group)
Rainfall	:	350 mm : 1700 mm annually
Main crops	:	All crops
Properties	:	These soils are sandy loam to clay loam in textured.

- Soil colour is light brown to dark brown.
- Soil pH is 7-8.
- Deficient in N,P and humus; moderate – high in K and excess in l
- Generally friable in nature
- Dominancy of illite mineral
- Sometimes, Kankar layers are found in 'B' horizon.

Black Cotton soils

The name black is given to soils that are very dark in colour and turn very hard on drying and sticky and plastic on wetting, and hence are very difficult to cultivate and manage. These soils are comparable with the Grumosols of the USA. In view of their black colour, they are comparable with the Chernozems of Russia and Prairie soils of the USA, but differ in their physico-chemical properties. These soils are dominantly distributed in the central, western and southern states of India like; part of Madhya Pradesh, Gujarat, Maharastra, West part of Andhra Pradesh, plateau of Deccan and Tamilnadu.

Order	:	Vertisols order
Area	:	72 M ha (IInd largest group)
Climate	:	Semi-arid to sub-humid climate
Main crops	:	Cotton, Bajra, Jowar, Ragi, Wheat and Millets

Properties

- Highly clay textured with clay content between 30-80%
- Soil colour is dark brown to black.
- Soil pH is 7.8- 9.5
- Deficient in N,P, OM and rich in K, Fe, Al, $CaCO_3$, $MgCO_3$, Montmorillonite
- Silica: sesquioxide ratio is 3 - 3.25
- Have high water holding capacity
- High sticky in nature
- CEC is 30-60 Cmol (P+)/kg
- Swelling and shrinking property occurs

Red soils

The name Red is given to soils rich in sesquioxides that have developed on rocks of Archean origin (granite, gneiss) and on well drained, stable and higher land forms under hot, semi-arid to humid sub-tropical climatic conditions. Under such conditions, the weathering is moderately intense and leads to enhanced decalcification. These soils are predominantly observed in whole part of Chennai, Mysore, part of Uttar Pradesh (Mirzapur, Hamirpur, Jhansi), Bihar (Santhal pargana, Chhotanagpur), West Bengal and eastern half part of Rajasthan, Madhya Pradesh, Andhra Pradesh and Orissa.

Order	:	Alfisols order
Area	:	70 M ha (3rd largest group)
Climate	:	Sub-humid climate
Main crops	:	Cotton, Bajra, Jowar, Ragi, Wheat and Pulses

Properties

- Loamy sand – clayey textured, poor in clay
- Soil colour is red due to presence of Hematite
- Neutral – acidic in nature
- Deficient in NPK and low in OM and lime
- CEC is < 20 Cmol (P+) / kg
- Silica: sesquioxide ratio is 2.

Laterite soils

Found in typical monsoon conditions-under conditions of high temperature and heavy rainfall with alternate wet and dry periods. The alterations of wet and dry season leads to the leaching away of siliceous matter and lime of the rocks and a soil rich in oxides of iron and aluminum compounds is left behind. These are found in Hill tops of south, hills of Rajasthan, plain part of Orissa, Maharastra, Malawar and Central India.

Order	:	Oxisols order
Area	:	25 M ha
Climate	:	Tropical and sub-tropical regions with high rainfall areas
Main crops	:	Low land-Paddy and Up land - Tea, Coffee, Rubber and Cinchona

Properties

- High clay content in soil
- Soil pH is 5-6.
- Essential plant nutrients are less.
- CEC is < 16 Cmol (P+)/ kg
- Silica: sesquioxide ratio is 1-1.25
- Characterized by compact, honey combed or vasicular structure in sub-soil.
- Deficient in lime

Desert soils

Location	:	North – West part of India i.e. Rajasthan, Kathiabar, south Punjab and in west district of Uttar Pradesh
Order	:	Aridisols order
Area	:	29 M ha
Climate	:	Arid regions
Main crops	:	Sorghum, Bajra, Pulses

Properties

- Sandy – loamy sand textured with clay content between 3.5 - < 10%
- Soil colour is light brown-yellowish brown.
- Soil pH is 7.8 – 9.2.
- Poor soil fertility status
- Organic matter content of the soils is very low.
- Poor water holding capacity
- High soluble salts

Tarai soils

Location	:	Tarai area of Punjab, Uttar Pradesh, Bihar, West Bengal and Himalaya
Order	:	Mollisols order
Area	:	5.7 M ha
Climate	:	Wet regime and high water table condition for most part of the year
Main crops	:	Sorghum, Bajra, Pulses

Properties

- Silty loam- loam textured soils
- Soil colour is black to grey black
- Soil pH is 4.7-5.8.
- Waterlogged soil has poor fertile
- Good drained soil has high fertile
- In lower horizons, stones and gravels are dominant

Saline and sodic soils

Location	:	Saline soils are found in Punjab, Uttar Pradesh., Rajasthan, Madhya Pradesh and Bihar, however Sodic soils are found in Gujarat, Maharastra and Karnataka.
Order	:	Aridisols order
Area	:	7 M ha
Climate	:	Arid to semi - arid climate
Main crops	:-	Paddy dominant

Properties

- Soil pH is > 7.3
- Poor soil fertility status
- Poor physical condition of soil
- In 'B' horizon, kankar layer is found.
- Soil colour is white – brown (saline) and black (sodic soil).
- Excess salts are present.
- Growth of most of the crop plant is reduced.

Organic soils

Location	:	Tarai of Himalaya, middle part of Bihar, N-E part of Uttar Pradesh, Kerala, Kochin, and Sundarban (West Bengal)
Order	:	Histosols order
Area	:	2720 km^2
Climate	:	Humid climate with low temperature
Main crops	:	Paddy dominant

Properties

- Soil colour is blue to black
- Soil pH is about 4
- OM is found 15 - 50% and sometimes up to 80%
- CEC is 50-100 Cmol (P+) / kg
- Soil horizons are very clear and mature.

Forest soils

The forest soils may be defined as a part of the earth surface which serves as a medium for the substances of forest vegetation. It is a living, dynamic, biological entity which is all the time undergoes changes in the course of forest vegetation and silvicultural systems. The formation of these soils is mainly governed by the characteristic deposition of organic matter derived from forest growth and found in Jammu & Kashmir, Uttarakhand, Himchal Pradesh, Uttar Pradesh, Bihar, Madhya Pradesh, Maharastra, Kerala and N-E regions.

Order	:	Deciduous forest - Histosols order
Evergreen forest	:	Spodosols order
Area	:	75 M ha
Dominant crops	:	Apple, Pear, Alucha

Characteristics

- These soils are characterized with the presence of litter with an associated micro flora and fauna.
- Root activities from the surface to the subsurface horizons are common phenomena in forest soils.
- Forest soil has natural well distinct horizons.
- Forest soils are characterized by higher porosity, higher permeability, more soil aggregates and excellent water holding capacity.
- In forest soil, the C/N ratio is higher than agricultural soil.
- Typical well developed forest soil profile consists of 4 horizons viz; Organic horizon (O), Alluvial horizon (A) zone of accumulation (B) and unaltered parent materials (C).

Podzol soils

Location	:	Himalaya mountain and in some part of south India
Order	:	Spodosols order
Area	:	Very less area in India (<1%)
Climate	:	Humid temperate region with high rainfall areas

Properties

- pH of upper most of soil is low and increases with increase in soil depth
- CEC of' 'A' horizon is less than the 'B' horizon.
- Si dominant in 'A' horizon and Al is less and vice versa in 'B' horizon.
- Good crops can not be taken due to absence of sufficient available plant nutrients.
- Sesquioxide and humus work as cement and produce hard layer and kankar in 'B' horizon.

Skeletal soils

Location	:	Laddhakh area of Himalaya mountain and Vindhyan region of central India
Order	:	Entisols order
Area	:	24 M ha
Climate	:	Cold climate

Properties

- Silty loam - loam textured soil
- Unfertile soil.
- Natural soil profile and horizons are absent.
- Soil colour is light yellow-brown.
- Very shallow soils (6-15 cm deep).

ORGANIC MATTER

Definition

The residue of animals, micro-organisms, plants, vegetation etc. which are found on the upper layer of soil surface or added from outside, is called organic matter.

Sources of organic matter

(i)	Plant sources	: Straw, Green leaves, Stem, Roots stubbles, Twings
(ii)	Animal sources	: Dead animals, Bones
(iii)	Other sources	: FYM, Compost, Green manuring, Fish meal, Dried blood etc.

Composition of plant residues:

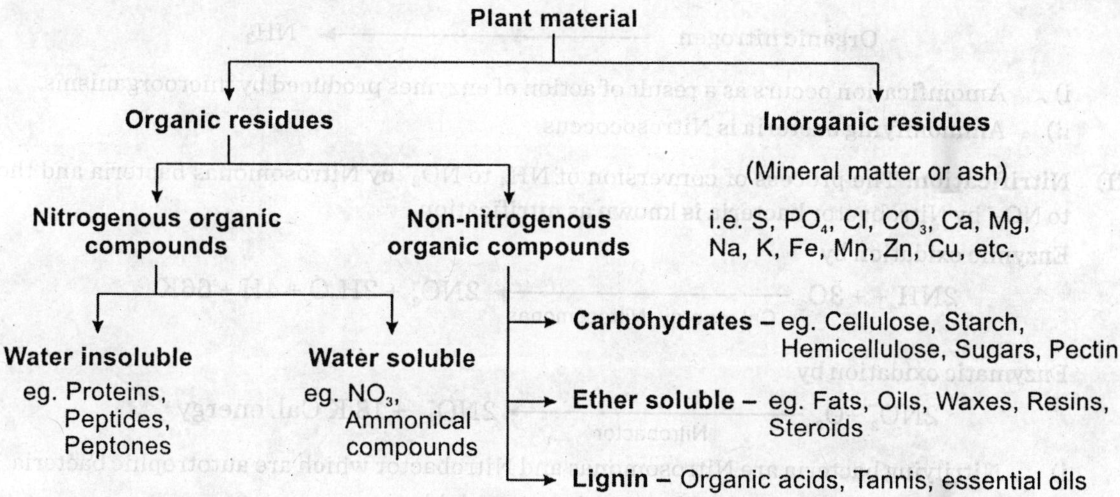

Green plant material

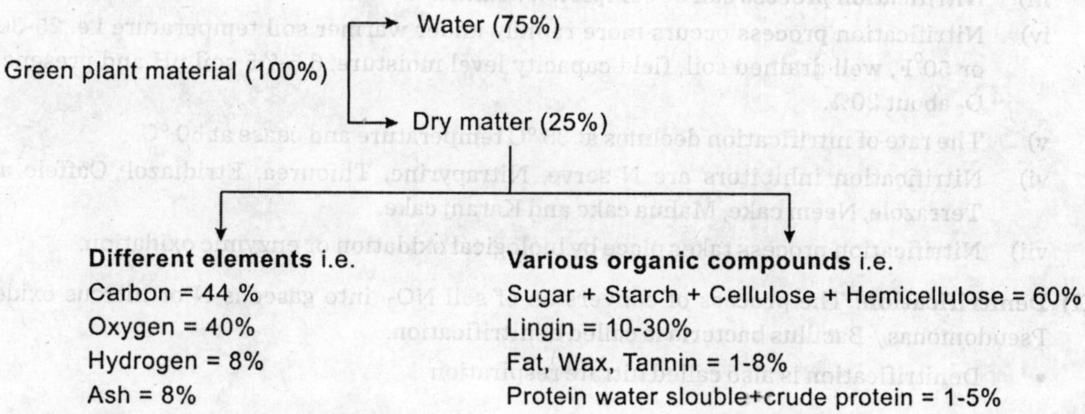

Decomposition of organic matter:

(A) Decomposition of soluble substances

Sugar and water soluble nitrogenous compound are the first to be decomposed and they produce very readily available sources of C, N and energy for the micro-organisms.

$$\text{Sugar (Carbon and Hydrogen containing substance)} + O_2 \xrightarrow{\text{Aerobic condition}} CO_2 + H_2O$$

$$\text{Sugar} + O_2 \xrightarrow[\text{nutrient and oxygen}]{\text{Restricted supply of}} \text{Aliphatic acids (Acetic, Formic acids) or Hydroxy acids (Citric, lactic acid) or Alcohals (ethyl alcohal, methyl alcohal)}$$

(I) Ammonification: The transformation of organic nitrogenous compounds into NH_4^+ forms by enzymatic hydrolysis through some intermediate steps.

Organic nitrogenous compounds → Polypeptides → Amino acids → NH_3 or salt of NH_4^+

or

$$\text{Organic nitrogen} \longrightarrow NH_3$$

i) Amonification occurs as a result of action of enzymes produced by microorganisms.

ii) Ammonifying bacteria is Nitrosococcus

(II) Nitrification: The process of conversion of NH_4 to NO_2^- by Nitrosomonas bacteria and then to NO_3^- by Nitrobactor bacteria is known as nitrification.

Enzymic oxidation by

$$2NH_4^+ + 3O_2 \xrightarrow[\text{Cal. energy Nitrosomonas}]{} 2NO_2^- + 2H_2O + 4H + 66K$$

Enzymatic oxidation by

$$2NO_2^- + O_2 \xrightarrow[\text{Nitrobactor}]{} 2NO_3^- + 18\text{ K Cal. energy}$$

i) Nitrifying bacteria are Nitrosomonas and Nitrobactor which are autotrophic bacteria.

ii) Nitrifying bacteria are highly sensitive to pesticides.

iii) Nitrification process can be completed within 2-4 weeks.

iv) Nitrification process occurs more rapidly under warmer soil temperature i.e. 25-30 °C or 50^{0}F, well drained soil, field capacity level moisture, 6.5-7.5 soil pH and presence of O_2 about 20%.

v) The rate of nitrification declines at 35 °C temperature and cease at 50 °C.

vi) Nitrification inhibitors are N-serve, Nitrapyrine, Thiourea, Etridiazol, Caffeic acid, Terrazole, Neem cake, Mahua cake and Karanj cake.

vii) Nitrification process takes place by biological oxidation or enzymic oxidation.

(III) Denitrification: The process of conversion of soil NO_3^- into gaseous N or nitrous oxide by Pseudomonas / Bacillus bacteria is called denitrification.

- Denitrification is also called nitrate respiration

$$2NO_3^- \xrightarrow{-2[O]} 2NO_2^- \xrightarrow{-2[O]} \uparrow 2NO \xrightarrow{-[O]} \uparrow N_2O \xrightarrow{-[O]} \uparrow N_2$$

Nitrate ions ($2NO_3^-$); Nitrite ions ($2NO_2^-$); Nitric oxide gas (N_2O); Nitrous oxide gas (N_2)

1. Denitrification occurs in rice field, poor drainage or under flooded conditions.
2. In denitrification, loss of N is as high as 87% of added N.
3. The rate of denitification is very low in acid soils with pH 3.6-4.8.
4. Denitrification occurs by reductase enzyme.
5. Denitrification occurs only if NO_3^- is present:

Control of denitrification

i. By improved drainage
ii. Deep placement of fertilizers
iii. Non nitrate fertilizers should be adopted
iv. Slow release N fertilizers may be used
v. Urea mud balls may be used
vi. Split application of fertilizers should be adopted
vii. Foliar application of Urea may be adopted

(B) Decomposition of insoluble substances

(i) **Breakdown of protein:** Proteins are complex organic substance and containing N, S, and some times P.

- Transported protein is Hemoglobin and Myglobin.
- Protein name was suggested by **Berjelius.**
- Proteins are the polymer of amino acid.

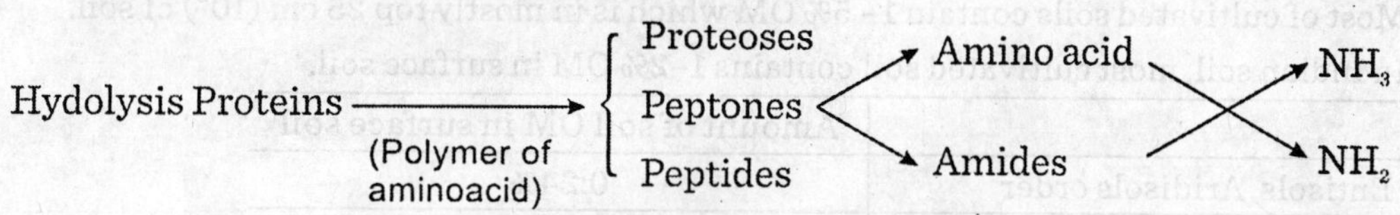

The conversion of proteins into amino acids, amines and Urea by the action of heterotrophic bacteria (*Rhizobium, Azotobactor, Clostridium, Amonifers, Denitrifers*) and fungi is known as **Aminization.** Bacteria in alkaline soil and neutral soils while fungi dominates in acidic soils. Amine ($R—NH_2$) is formed by **Aminization.**

Factors affecting availability of organic matter (OM)

1. **Climate:**

   i) In cold climate: Richer OM due to slow decomposition of OM.
   ii) With increase rainfall: Rich OM due to more growth of vegetation.
   iii) In warmer climate: Poor OM due to fast decomposition of OM.

2. **Natural vegetation:**

Grass land areas
Trees areas } more OM

Note: Organic matter content of grass land soil is generally higher than forest soils mainly in surface soil.

3. **Texture:**
   Light textured soil – Less OM due to fast decomposition of OM
   Heavy textured soil – More OM due to slow decomposition
4. **Structure:**
   Spheroidal structure – Rich OM due to good soil physical condition
5. **Drainage:**
   Poorly drained soil – High OM due to poor aeration
   Good drained soil – Less OM due to good aeration
6. **pH:**
   Best pH – 6-8
7. **Moisture:**
   Dry and waterlogging soil condition – Slow decomposition of OM
8. **Tillage:**
   Continuous tillage – Poor OM
9. **Cropping system:**
   Cultivated soils – Decrease OM

Legumes are more readily decomposed than grasses.

- Most of cultivated soils contain 1– 5% OM which is in mostly top 25 cm (10") of soil.
- In Indian soil, most cultivated soil contains 1–2% OM in surface soil.

	Amount of soil OM in surface soil
Entisols, Aridisols order	0.24%
Rajasthan soil	0.33%
Alluvial soil	0.33–0.67%
Deep black soil	0.68–6.53%
North Karnataka and AP	1.37%
Annual rainfall < 500 mm	0.59%
Annual rainfall 500-1000 mm	1.20%
Annual rainfall > 1000 mm	0.45–1.55%

- Organic matter content of most Indian soils is < 1%

Carbon : Nitrogen ratio (C:N ratio)

Definition: The ratio of weight of organic carbon (C) to the weight of total nitrogen (N) in a soil or OM is known as C:N ratio. The importance of C:N ratio is in mineralization and conserve N and OM in soil.

Organic matter/soil	C : N ratio
Soil microbes	4:1-9:1
Bacteria	5:1

Organic matter/soil		C : N ratio
Actinomycates	:	6:1
Fungi	:	10:1
Humus	:	10–12:1
Green legumes	:	12:1
Alfalfa	:	13:1
Clovers	:	20:1
Manures	:	20:1
FYM	:	20–30:1
Berseem	:	21:1
Hay (Legume grasses)	:	25:1
Blue grass	:	30:1
Green rye	:	36:1
Green grasses	:	40:1
Maize stalk, Sugarcane Bagashi	:	40–50:1
Grain straw, Dry grasses	:	80:1
Rice stover	:	89:1
Wheat straw	:	124:1
Jowar stalk	:	93:1
Wood burada	:	350:1
Saw dust	:	400–500:1
General soil, Cultivated soil /arable land	:	8–15:1
Mineral soil, Waterlogged soil	:	10–12:1
Organic matter (OM)	:	40:1–12:1
Tropical soils	:	10:1

Important points

1. If C:N ratio of OM is < 20, resulting more N mineralization and more gain of available N in soil.
2. If C:N ratio of OM is 20–30, resulting no gain of available N.
3. If C:N ratio of OM is > 30, resulting less N-mineralization and more N- immobilization and finally loss of available N in soil.
4. If decomposition of manure is slowly, then wider/higher C:N ratio.
5. If decomposition of manure is fastly, then lower/ narrow C:N ratio.
6. C:N ratio of sub surface soil is narrow than surface soil.
7. C:N ratio of dry areas soil is narrower than humid areas soil.
8. C:N ratio of mature residues i.e. legumes or non-legumes is higher than younger succulent materials.
9. C:N ratio of warmer regions (arid climate) is lower than colder regions (humid climate).
10. As decomposition occurs, C:N ratio decreases due to C is lost and N is conserved.
11. C:N ratio of less carbohydrate in OM is narrow/ lower than the high carbohydrate in OM.

12. C:N ratio of Indian soils is 14:1.
13. Lower C:N ratio means greater amount of mineralization of N to NH_4^+ and NO_3^- form.

Maintenance of soil OM

1. Minimum tillage
2. Growing of grasses in arid and semi arid regions - less water requirement of grasses.
3. Growing green manuring crops - High N content (2.3–3%).
4. Addition of crop residues (Mulch).
5. Use of organic manures i.e. Compost, FYM, Urban and Industrial waste.

Role of organic matter

1. Makes good physical condition of soil.
2. Makes P and K more available in acidic soil.
3. Maintains favourable condition of aeration, infiltration and permeability.
4. Maintains optimum temperature.
5. Increases WHC and buffering capacity of soil.
6. Decreases water erosion, wind erosion and alkalinity of soil.
7. Prevents leaching loss.
8. Helps in growth of micro-organisms
9. Supplies food for micro-organisms.
10. Serves as reservoir of mineral nutrients.

Organisms involved in mineralization of N

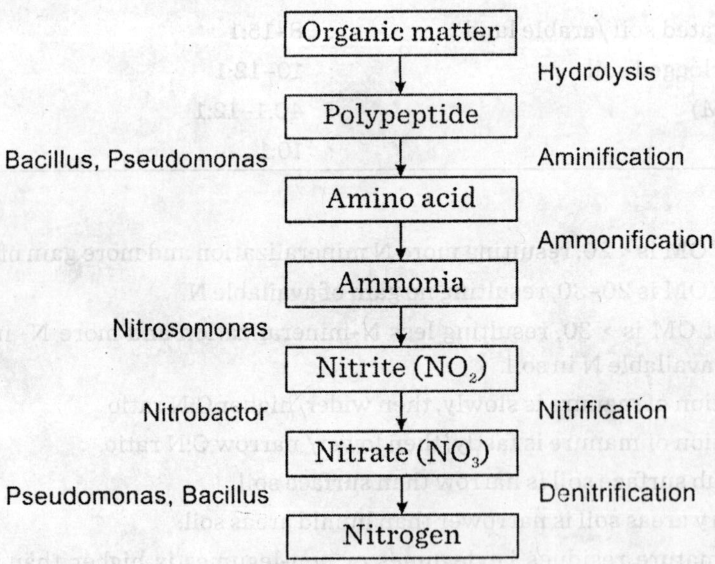

Key points

1. Percentage of P in OM = 0.5%
   Percentage of N in OM = 5%
   Percentage of S in OM = 0.5%

2. **N-mineralization: The conversion of organic N to NH_4^+ and NO_3^- is known as N-mineralization. Mineralization of organic N involves three reactions:**

(a) Aminization
(b) Ammonification
(c) Nitrification

} Occurs by the activity of heterotrophic micro-organism nism

3. **N-Immobilization:** The conversion of inorganic N to organic N is known as N-immobilization.
   - i It is opposite process of N- mineralization.
   - ii If decomposing organic matter contains low N than C, micro-organism will immobilize inorganic N in the soil.

4. **Volatilization:** NH_3 volatilization refers to the loss of N from the soil.

Facts:

i. NH_3 volatilization capacity are graded as:

Urea > $(NH_4)_2SO_4$ > NH_4Cl > NH_4NO_3

Calcareous soil >Alkaline soil > Neutral soil > Acidic soil

ii. In neutral and acidic soils, NH_3 volatilization occurs from Urea only.

iii. Surface application of fertilizers increases NH_3 volatilization.

iv. NH_3 volatilization from bare soil, coarse textured soil, calcareous soil or alkaline soil is large.

Management

i. Urea should be never left on the soil surface without incorporation.

ii. In coarse textured soil, Urea may be applied on the surface after irrigation at suitable moisture.

iii. If soils are highly alkaline or highly calcareous, then deep incorporation of Urea should be made.

iv. Foliar application of Urea may be adopted.

v. Split application of Urea may be adopted.

vi. Urease inhibitors:

vii. Hydroxamic acid, Quinone, Acetohydroxymate, Thio urea, Thiocarbamate, Methyl urea

Humus

Definition: Humus is defined as a complex and generally resistant mixture of brown or dark brown, amorphous and colloidal material modified from the original tissues or synthesized by the various soil organisms.

(1) It is a natural body having variable and heterogeneous mass.

(2) Humus is a OM but all OM are not humus.

(3) Humus particles are colloidal in nature.

Classification of Humus

(A) According to **Molder,** humus is classified into 3 groups on the basis of colour and solubility in water and alkali solution.

(i) Soluble in water : eg. Crenic, Apocrenic acid

(ii) Soluble in alkali : eg. Almic (brown), Humic acid (black)

(iii) Insoluble in alkali : eg. Almin, Humin

(B) According to **Oden**, humus is classified as below

(i)	Humus coal	(iii)	Humic acid
(ii	Hematomelanic acid	(iv)	Fulvic acid

Composition of humus

On weight basis

Lignin	:	55%	Carbon	:	50-60%
Amino acid	:	35%	Hydrogen	:	3-4%
Carbohydrate	:	15%	Oxygen	:	30-40%
Hemicellullose	:	7%	N	:	4-5%
Resin, fat, wax	:	3%	P	:	0.4-0.5%
Others	:	5%			

Properties of Humus

1. Humus is fairly stable.
2. The colour of humus is brown, dark brown to black.
3. The C:N ratio of humus is 10-12:1.
4. Humus is resistant to decomposition.
5. Humus is formed by C, H, O, N, P, S.
6. Humus is a colloidy and non-crystalline in nature.
7. CEC of humus is 100-300 Cmol(P^+) / kg soil.
8. Humus absorbs 80-90% water under saturated condition.
9. Adsorption power and surface area are very higher than silicate clay.
10. The density of humus is between 1.3-1.5 Mg/m^3.
11. Colloidal surfaces of humus are negative charged.
12. Humus is hygroscopic and acidic in nature.
13. Humus has swelling capacity in presence of water.
14. Humus increases adhesion, cohesion and plasticity properties in sandy and sandy loam soils but these properties decrease in clayey soils i.e. sandy clay, silty clay and clay.
15. Humus is practically insoluble in water but soluble in weak acid solution and dilute in alkali solution i.e. NaOH, KOH.
16. Carbon content of humus is less variable than N.

Importance of Humus:

1. Humus makes good physical condition of soil.
2. Humus makes P and K more available in acidic soil.
3. Humus maintains good condition of aeration, infiltration and permeability.
4. Humus increases WHC, buffering capacity, soil colour and CEC of the soil.
5. Humus decreases water erosion and wind erosion.

6. Humus reduces evaporation, alkalinity of the soil and leaching losses.
7. Humus helps in growth of micro-organisms and supplies food for micro and macro organisms.
8. Humus serves as reservoir of chemical elements.
9. Humus reduces soil temperature in summer and maintain hot in winter season.
10. Humus acts as chelate and buffering agent.
11. Humus neutralizes substances which have toxic effect on plants.

Problem soils

A. Salt affected/Saline soils

Soils of arid and semi-arid regions contain high concentration of either soluble salts or exchangeable Na or both due to poor leaching of base forming cations. The soil pH is generally >7 and soil colloidal complex dominated by Ca^{++}, Mg^{++}, K^{+} and Na^{+} ions.

Classification

On the basis of soil pH, SAR, ESP and EC, salt affected soils are classified into 4 groups:

- Saline soil
- Non-saline alkali soil
- Saline- alkali soil
- Degraded alkali soil

Particulars	Saline soil	Non-saline alkali soil	Saline - alkali soil	Degraded alkali soil
Other name	American scientist **E.W. Hillgaurd** gave the name White alkali, Russian scientist **Gaidraize** gave the name solanchak, Hungory scientist gave the name Szek and in India, their local name is Reh, Kari, Kallar, Khar, Pokkati, Shora, Soulu, Loni, Thur (Punjab), Brown alkali soil	Black alkali Solonetz Szik and in India, their local name is Black alkali typical usar, Alkali soil, Sodic soil, Chaupan soil (Bombay), Chaundu soil, Kshar soil, Rakkar, Bara-bari soil (Punjab)	USAR soil	Solod soil, Soloth soil
Soil pH	< 8.5	8.5-10	> 8.5	6 (surface soil), 8.5 (sub surface)
SAR	< 13	> 13	> 13	<13 (surface), >13 (sub surface)
ESP	< 15	> 15	> 15	> 15

Particulars	Saline soil	Non-saline alkali soil	Saline - alkali soil	Degraded alkali soil
EC of saturated extract (dSm^{-1})	> 4	< 4	> 4	< 4
Soluble salts (%)	> 1%	< 1%	> 1%	< 1%
Dominant salts	SO_4^{--}, Cl^-, NO_3^- of Na, Ca, Mg and K CO_3^{--} and HCO_3^- are also found in small amount	Na_2CO_3	-	Na_2CO_3 in lower layer
Content in soil	High soluble salts of Na	High exchangeable Na on the soil colloidal complex	Na clay and soluble salts	H^+ in upper layer and Na+ in lower layer
Soil colour	White due to SO_4^{--} and Cl^- of Na, brown due to $NaNO_3$	Black colour due to Na_2CO_3 content	-	Black colour in lower layer due to Na_2CO_3
OM content	Slightly less than normal soil	Very less amount due to dispersion of OM	Low	Low
Climate	Arid and semi-arid regions, Rainfall < 550 mm annually	Semi-arid & Sub humid region, Rainfall > 550 mm annually	Semi-arid & Sub humid region	High rainfall areas
Forms by the process	Salinization	Alkalinization	Salinization, Alkalinization	Solodification or solodization
Physical condition of soil	1. Flocculated or granulated condition 2. Good permeable to air and water 3. Soil structure is optimum	1. Deflocculated or un-granulated condition 2. Poor permeable to air and water 3. Soil structure is very poor	Flocculated and Deflocculated condition	1. Low rate of infiltration and permeability 2. It develops columnar or prismatic structure 3. Compacted soil
Recommended fertilizer	$(NH_4)_2SO_4$, CAN, SSP, TSP	$(NH_4)_2SO_4$, NH_4NO_3, DAP	NH_4NO_3, DAP, TSP	Basic fertilizes i.e. $Ca(NO_3)_2$, $NaNO_3$
Harmful effect	1. Available moisture is less 2. Poor seed germination due to poor aeration	1. Dispersion of soil particles 2. Poor soil physical condition 3. Na generated Ca and Mg deficiency	1. Poor internal drainage due to hard clay pan formed by	Very low P and due to leaching of P

Particulars	Saline soil	Non-saline alkali soil	Saline - alkali soil	Degraded alkali soil
	3. Soils are generally barren 4. Wilting coefficient is very high 5. Soil mass is structureless and dusty 6. Conc. of K is low 7. Conc. of B, F, NO_3 are high which affect the plant growth 8. Toxic effect of CO_3^{--}, Cl^-, SO_4^{--}, of Na 9. Root injury 10. Leaf tip burnt / chlorosis	4. N, Zn, Fe, Cu are low 5. P, K, B are high 6. Toxicity due to high pH, CO_3^{--} or HCO_3^- of Na, and B 7. Crust formation during dry 8. Sticky during wet	dispersed Na-clay 2. Poor plant growth	

Particulars	Saline soil	Sodic soil
Surface soil under dry condition	Friable soil surface	Hard surface
Sub surface hard pan	Sometimes thin hard pan	Generally thick hard pan
Collected water in surface	Clean	Muddy
Collected water in sub surface	Bad	Good
Absorption of rain water or irrigation water	Soon	Waterlogged on surface soil due presence of hard pan

Order :

1. Salt affected soils are grouped under Aridisols order according to 7th Approximation (C.E. Kellogg, M. Baldwin and J. Thorp, 1960).
2. According to Thorp and Smith (1949), salt affected soils are grouped under Intrazonal order or Halomorphic sub-order (Genetic system of soil classification).

Area:

Area (M ha)	Location
957 M ha	In world
7 M ha (Abrol and Bhumbla, 1971)	In India
10 M ha (Bhargava, 1989)	In India
13 M ha (Vohra, 1987)	In India
1.295 or 1.3 M ha	In U.P.

Sodic soil (3.74 mha) : Haryana, Gujarat, Maharashtra, Coastal areas
Saline soils (3.0 mha) : Rajasthan, Punjab, U.P., Bihar
Uttar Pradesh : Aligarh, Mainpuri, Etah, Etawah, Kanpur, Unnao, Raibarely, Hardoi, Pratapgarh, Jaunpur, Farrukhabad, Agra, Mathura, Meerut, Fatehpur, Allahabad, Sultanpur districts.

FORMATION OF SALINE AND ALKALI SOILS

According to **D. Sigmond** (1932), the saline and sodic soils are developed by 5 stages.

1. Salinization

The process of accumulation of soluble salts in soil is known as salinization and the soil so formed is called saline soil.

Source of soluble salts

1. Arid and semi arid regions where low rainfall and high evaporation occurs.
2. Irrigation by more salty water up to long period
3. Poor water drainage or restricted water drainage
4. Very high water table
5. Presence of hard layer in soil profile
6. Nature of parent material
   Basic rocks (45-55% Silica) = eg. Basalt, Gabbro
   Sub-basic rocks (55-60% Silica) = eg. Diorite, Andesite
7. More use of basic fertilizer up to long period
   $Ca(NO_3)_2$ = 15.5% N + 19.5% Ca + 1.5% Mg
   $NaNO_3$ = 16% N + 26% Na
   Basic slage $[(CaO)_5.P_2O_5.SiO_2]$ = 14 - 18% P_2O_5 + 15 - 34% Ca + 1.8 - 9% Mg
   Rock phosphate $[Ca_3(PO_4)_2]_3CaCO_3$ = 20 - 40% P_2O_5 + 35 - 38% Ca
   Calcium cynamide $[CaCN_2]$ = 21% N + 39.1% Ca
8. Flooding by sea water
9. Ploughing in same depth up to long period
10. Slow rate of permeability of soil
11. Basin shaped topography

2. Alkalinization

The accumulation of Na+ ions on soil colloidal complex is known as alkalinization and the soil so formed is called alkali or sodic soil.

OR

Na+ replaces adsorbed exchangeable cations i.e. Ca++, Mg++, and K+ from soil colloidal complex by base exchange process and then Na+ adsorb in place of replacing cations, resulting Na+ increases on soil colloidal complex, the process is called alkalinization and the soil so formed is called alkali or sodic soil.

3. Desalinization

Movement or removal of soluble salts out of a section of the soil profile is known as desalinization and the soil so formed is called saline alkali soil.

OR

Soluble salts goes down in lower horizons by leaching from the upper soil surface but adsorbed Na+ remain on soil colloidal complex, due to high amount of Na+ on soil colloidal complex are dispersed, the process is known as desalinization.

4. Degradation or De-alkalization/ Solodization:

Solodization refers to the removal of Na+ from the exchange sites and out of section of profile.

OR

Leaching of Na+ and salts from surface soil and pH is about 6, the process is known as degradation and the soil so formed is called degraded alkali soil.

OR

In this stage, the exchangeable Na+ ions are hydrolyzed and form NaOH and H+ .NaOH reacts with CO_2 present in soil and form Na_2CO_3. Na_2CO_3 is more soluble in water and goes down in lower horizons by leaching. H+ replaces adsorbed Na+ from soil colloidal complex, resulting H+ increases on soil colloidal complex and finally surface soil pH is about 6, the process is known as degradation and the soil so developed is called degraded alkali soil.

5. Regradation

The soil in which the process of degradation has occurred and bases are to be re-accumulate and amount of Na salts increased, the process is known as regradation.

Effect of salinity and alkalinity on soil and plants

On soil
- Poor soil structure
- Poor aeration
- Slow permeability rate
- Slow bacterial activity
- Pudding of soil
- Difficulty in agricultural works
- Nutrient imbalance
- Poor physical condition of soil
- Deficiency of Fe, Mn, Zn, Cu
- Oxygen deficiency

On plants
- Hindrance in water absorption
- Poor uptake of Ca^{++} ; Mg^{++} by plants
- Toxic effects of individual ions

Reclamation and management of saline soil

Following methods may be used for reclamation and management of saline soil:

(A) Physical measures or mechanical measures

1. Scrapping of surface soil : 15-30 cm depth, suitable for patchy field.

2. Leaching of soluble salts below the plant root zone.

$$\text{Leaching requirement (\%)} = \frac{\text{EC of applied water}}{\text{EC of drainage water}} \times 100$$

OR

$$\text{Leaching requirement (LR)} = \frac{EC_{lw}}{EC_{dw}}$$

- Leaching process is more useful in salt affected soil where water table is down.
- Summer season is more suitable for leaching.

3. Flushing of soluble salts:
   Removal of soluble salts from soil surface by runoff from upper portion to lower portion is called flushing. This process is more useful for salt affected areas where water table is high and soils having gentle slope.
4. Trenching:
   - 50 cm × 50 cm × 50 cm one trench and again after 2m distance.
   - Suitable where sub soil is good.
5. Washing out of soluble salts from surface layer: More suitable where water is cheap and excess amount.
6. Proper water drainage

(B) Agronomical practices or cultural practices

1. Deep ploughing with the help of soil turning plough i.e. Victory plough to open hard layer.
2. Growing of salt tolerant crops.

High salt tolerant crops (> 5000 ppm salts)	Medium salt tolerant crops (2500-5000 ppm salts)	Low salt tolerant crops (< 2500 ppm salts)
Barley, Cotton, Dhaincha, Lucern, Rice (transplanted), Sugarbeat, Senji, Safflower, Toria, Tobacco, Turnip, Taramira, Karnal grass, Date palm, Ber etc.	Arhar, Berseem, Bajra, Carrot, Cauliflower, Cabbage, Pea, Paddy (direct sown), Wheat, Oat, Jowar, Maize, Mustard, Alfalfa, Tomato, Guava, Sweet pea, Fig etc.	Moong, Urd, Cow pea, Lentil, Gram, Groundnut, Beans, Radish, Onion, Apple, Pear, Plum, Peach, Grape, Orange, Lemon etc.

3. Suitable crop rotation: Dhaincha – Paddy – Berseem
   Dhaincha – Wheat
   Paddy – Barley – Cowpea
   Paddy – Mustard
   Gaur – Barley
4. Increased plant population / unit area.
5. Sowing of seeds in furrow.
6. Use of acidic nature fertilizers i.e.
   $(NH4)_2SO_4$: 20.6% N + 24% S
   NH_4Cl : 26% N + 66.3 % Cl

7. Salt free water should be used for irrigation purpose.
8. Light irrigation is better than heavy irrigation.
9. Sprinkler irrigation is better.
10. Application of organic residue mulch @ 4 t ha^{-1} on surface soil to reduce the evaporation rate.

(C) Biological measures

1. Application of FYM @ 50 t ha^{-1} : 0.5% N + 0.25% P_2O_5 + 0.5% K_2O
2. Growing of green manuring crops i.e. : Dhaincha (*Sesbania acculeata*)
   Dhaincha is more suitable for green manuring because
   i. High salt tolerant crop
   ii. Vigorous root system
   iii. Fast growth
   iv. More biomass
3. Use of molasses i.e. Sira, pressmud
4. Use of 2.5 t ha^{-1} of dry powder of Satyanasi weed (*Argemone maxicana*)
5. Use of Usartor Masala : It is a organic amendment

Reclamation and management of sodic soil

Main aim of reclamation of sodic soil is to replace exchangeable Na with Ca.

Important points: Of all Ca compounds, gypsum is considered best and cheapest for reclamation of sodic soil because

1. High soluble Ca salt
2. Easily available in the market
3. More amount available in India

B. Gypsum:

Chemical formula : $CaSO_4.2H_2O$

Composition : 23.2% Ca +18.6 -22.5% S +20.9 % water

Gypsum requirement:

The amount of gypsum required to be added in a sodic soil to lower the ESP to a desired value is known as gypsum requirement. It is expressed as me of Ca /100 g soil.

$$\text{GR (me of Ca / 100 g soil)} = \frac{\text{CEC} \times [\text{ESP (initial)} - \text{ESP (final)}]}{\text{CEC}} = 100$$

Where, ESP (initial) = ESP value before application of gypsum

ESP (final) = It is kept at 10

CEC = Cation exchange capacity – Cmol(p^+)/kg soil

Important points:

1. Common method used for determination of GR of sodic soil is Schoonover's method (1952).
2. Amount of gypsum required for reclaiming sodic soil is governing by;
   i Degree of sodicity

ii Crops to be grown

iii Soil texture

iv Drainage condition and quality of irrigation water

Method :

Agricultural grade with 70% pure gypsum should be powdered (sieving with 2 mm size sieve) with mixed thoroughly 10-12cm soil depth with shallow tillage followed by planking. About 25 t / ha of FYM should be applied. The field should be irrigated and kept ponded for 15 days with 6-8 cm standing water.

B. Acidic soils

Definition: Soils in which acidity of adsorbed H^+ ions on the colloidal complex is more than bases and pH is < 7 are called acid soils. In acid soils, most of the plant nutrients are unavailable and resulting plant growth is affected. Area of acid soils in India is about 49 M ha , out of which 26 M ha of land having the pH < 5.6 and 23 M ha of land having pH ranging from 5.6 - 6.5.

Distribution:

1. Acid soils occur generally in all major soil groups except black cotton soil.
2. In India , acid soils are found in North-Eastern and peninsular regions (including Bihar plateau and sub plateau of West Bengal and major part of the Orissa) and Himalyan regions (comprising Jammu and Kashmir).
3. There are large area of acidic soils in Assam, Manipur and Tripura.

Characteristics:

1. Low in soil pH and high in exchangeable hydrogen and aluminum ions are the main properties of acid soils.
2. These soils have low CEC and high base unsaturation.
3. These soils are low in available P and have high phosphate fixation capacity.
4. The status of available micronutrients except Mo is generally enough in these soils.
5. Soil acidity checks biological N-fixation.
6. The population of bacteria and actinomycetes are lower and fungi are higher in acidic soil than normal soil.
7. Soil acidity changes number and activity of micro-organisms.
8. Deficiency of Ca and Mg in these soils
9. Nutrient imbalance

Classification of acid soils:

Strongly acidic soils : pH of these soils are < 4

Moderately acidic soils : pH of these soils are > 4 - < 6

Slightly acidic soils : pH of these soils are > 6 - < 7

Acid soils in India :

- Laterite soil
- Lateritic soil
- Peat soil
- Muck soil
- Foot hill soil
- Podzolic brown soil
- Mixed red, black and yellow soil

Effect of soil acidity or problems of soil acidity:

1. Soil acidity increases the concentration of Fe and Al in soil solution resulting toxicity to plants.
2. Availability of micronutrients i.e. Fe, Mn, Zn, Cu, Cl, B increases, resulting poor plant growth.
3. Availability of Ca and Mg reduces
4. P fixation occurs in acid soils resulting deficiency of available P which affect the roots growth.
5. Beneficial activities of soil micro-organisms adversely affected.
6. Bad effect on plant phloem
7. Bad effect on activity of enzymes related to root and stem
8. Toxic effect of H^+ ions on root tissues
9. Various diseases occur in plants i.e.
   - "Finger toe disease" in Cruciferae family plants.
   - Root-rot of tobacco
   - Blight of potato
   - Root disease in mustard

Reclamation and management of acid soils

Following methods can be used for reclamation and management of acidic soils:

(A) Addition of liming material

(B) Growing of acid tolerant crops

(C) Use of basic fertilizers

(D) Soil and water management

Important points:

- To a chemist, lime is a CaO but to a farmer, agronomist and soil scientist, lime means $CaCO_3$ or $CaCO_3$ equivalent.
- Sequence of causticity: Hydrated lime > Quick lime < Limestone

2. **Effective Calcium Carbonate (ECC) or Neutralizing Index (NI):**

It is defined as the product purity of $CaCO_3$ equivalent and fineness factor. It is expressed as percentage.

3. **Calcium Carbonate Equivalent (CCE) or Neutralizing Value (NV) or Total Neutralizing Power (TNP):**

CCE is defined as the acid neutralizing capacity of the agricultural liming material and expressed as the weight percentage of $CaCO_3$.

$$CCE = \frac{\text{Molecular weight of pure } CaCO_3}{\text{Molecular weight of a liming material}} \times 100$$

whose CCE is to be determined

Liming materials	CCE(%)
$CaCO_3$	100
CaO	179

Liming materials	CCE(%)
$Ca(OH)_2$	135
$CaCO_3.MgCO_3$	108.7
$CaSiO_3$	86

4. Lime requirement:

Amount of liming material required to be added to an acid soil to increase the soil pH to a desired value is known as lime requirement. The method of lime requirement in acid soil was given by H.E., Shoemaker, E.O. Mc Lean, and P.F. Pratt (1961).

Lime requirement depends on following factors:

i) **pH of soil:** More lime is needed for strongly acidic soil than slightly acidic soil.

ii) **Soil texture:** More lime is needed for heavy textured soil than light textured soil.

iii) **Organic matter:** More lime is needed for organic soil due to more CEC and buffering capacity than mineral soil.

iv) **Type of clay minerals:** More lime is needed for Montmorillonite type of clay mineral due to high CEC and buffering capacity than Kaolinite and Illite clay mineral present in soil.

v) **Buffering capacity (BC):** Higher BC of soil, the greater amount of lime applied to attain satisfactory change in soil pH.

vi) Pulses and legumes require large amount of lime.

Note: Lime recommendation is made on the basis of a 15 cm furrow slice.

5. Method of lime application:

The most efficient way to use lime is to apply small amounts in each year or alternate year. Generally, it is applied one or two months before the sowing of crop. Lime should be powdered (sieved by 60 mesh sieve) with mixed thoroughly 10cm soil depth with shallow tillage followed by planking.

1. Effect of lime on soil:

a. Physical effects:

i) Improve soil structure

ii) Bulk density reduced

iii) Increase in infiltration and permeability rate

iv) Indirect help in control of soil erosion

b. Chemical effects:

i) Increases soil pH

ii) Decrease H^+ ions concentration and increase OH^- concentration.

iii) Reduction in toxic effect of heavy metal of Fe, Al, Mn.

iv) Lime supplies Ca and Mg for plant growth.

v) Lime increases the availability of N by the decomposition of OM

vi) Lime makes P and K more available.

c. **Biological effects:**

i) Fast decomposition of OM

ii) Increases activity of aerobic micro-organisms.

A. **Growing of acid tolerant crops:**

(i) **Very strongly acid tolerant crops:** Blue berry, Cranberry, Pine apple

(ii) **Strongly acid tolerant crops:** Rice, Potato, Sweet potato, Oat, Caster, Rai, Sorghum, Bajra

(iii) **Moderately acid tolerant crops:** Wheat, Barley, Maize, Turnip, Soybean, Cotton, Cow pea

(iv) **Slightly acid tolerant crops:** Alfalfa, Beans, Carrot, Cabbage, Cauliflower, Red clover, Sweet clover, Tomato.

B. **Soil and water management:**

Proper soil and water management check the leaching of bases and increase the decomposition of organic matter.

C. **Use of basic fertilizers:**

Sodium nitrate ($NaNO_3$), Calcium nitrate [$Ca(NO_3)_2$], Calcium cynamide ($CaCN_2$), CAN, Basic slage [$(CaO)_5.P_2O_5.SiO_2$], and Rock phosphate [$Ca_3(PO_4)_2]_3.CaCO_3$.

D. **Waterlogged soils**

Definition: Waterlogged or wet land can be defined as the soils whose development and properties are strongly influenced by temporary or permanent saturation in the upper part of the soils.

The soil is intermittently saturated with water, i.e. oxidation of OM is slow and it accumulation in the 'A' horizon. In the second horizon, Fe and Mn are deposited as rusty mottles, if the diffusion of O_2 into the soil is low, if the diffusion rapid, they are deposited as concretions.

Type of waterlogging:

Waterlogging is of following types:

1. **Riverine flood waterlogging:** In the rainy season, flood may be come to the nearby land from the river having excess flood water.
2. **Oceanic flood waterlogging:** Ocean water spreads in the nearby land caused by waterlogging.
3. **Seasonal waterlogging:** Runoff water accumulates in the low lands in the rainy seasons.
4. **Perennial waterlogging:** Deep water, swamp *etc.* get rain water, runoff water and seepage water from canal causing perennial waterlogging.
5. **Sub-soil waterlogging:** High water table in the rainy season is normally unsuitable for root-growth.

Factor affecting formation of waterlogged soils

(i) **Climatological factor:** Due to high rainfall, water accumulates on the soil surface.

(ii) **Flood :** Flood water is generally causing waterlogged situation in the field.

(iii) **Seepage from canal:** Ground water levels are closer to the surface due to seepage from canal.

(iv) **Land shape :** Saucer shaped land gets water from highland which results into waterlogging.

  (i) **Uncontrolled and unwanted irrigation:** Excess irrigation may cause accumulation of water on the soil surface.

  (ii) **Drainage:** Poor drainage system is responsible for waterlogging in the field.

Characterization of waterlogged soils

The following are the classes of depth to the wet state:

Class 1. Not wet above a depth of 400 cm
Class 2. Not wet above a depth of 100 cm
Class 3. Not wet above a depth of 50 cm
Class 4. Not wet above a depth of 25 cm

Effect of water logging:

1. Poor aeration
2. Poor structure
3. Makes compact soil
4. Lower down soil temperature
5. pH tends to increase in a acidic soil and decrease in sodic soil
6. All crops except paddy cannot be survive
7. Oxidation of organic matter is very slow and it accumulates in the 'A' horizon.
8. In 'B' horizon, Fe, Al and Mn are deposited as rusty mottles, if diffusion of O_2 is slow.
9. If diffusion is rapid, they deposited as concretion
10. Availability of nutrients

Management of waterlogged soils:

(i) **Leveling of land:** Leveling of land in many wetlands removes water by runoff.

(ii) **Water drainage:** Drainage removes excess water from root zone that is harmful for plant growth. Land can be drained by surface drainage, sub-surface drainage and drainage well method.

(iii) To check the seepage from the canals and irrigation channels

(iv) **Flood control measures:** Construction of bund may check water flow from the rivers to the cultivated lands.

(v) **Plantation of trees having high transpiration rate:** Transpiration rate in certain trees like; Eucalyptus spp, Accacia spp, Zizyphus and Salix *spp.* are very high. In transpiration process, the underground water is consumed by trees, thus lowering the ground water table.

(vi) Selection of crops and their proper varieties: Waterlogging tolerant crops and their varieties

Situation of water logging	Crop	Varieties
Deep water	Rice	Sudha, Janki, Jaladhi, Madhukar, Jalmagna, BR-14
Shallow water	Rice	Pankaj, Radha, Vaidehi, Rajshree

Waterlogging for short period	Rice	BR-13, Janki
Water table near the surface and sometimes waterlogging	Rice Sugarcane Sorghum Bajra Jute	Mahsuri, Jaishree, Sita, Kanak, Sujata BO-3, BO-29 T85 T55 JRC-212, JRO-632

(vii) **Sowing on the bunds or ridges:** In the waterlogged area, sowing should be done on bunds or ridges. In this method, there is scope of good aeration near the root zone of the crops.

(viii) **Nutrient management:** Low N fertility is an important constraint in the waterlogged area. The predominant form of inorganic N in low land soils is NH_4^+. The following concept may increase N- use efficiency.

1) Deep placement of fertilizer
2) Use slow and controlled release fertilizers
3) Use of nitrification inhibitors – i.e. Nitrosomonas bacteria
4) Use of urease inhibitors – i.e. Phenyl phosphorodiamidate.

Distribution: Certain areas in high rainfall zones where drainage is poor become waterlogged, generally in depressions and along rivers and streams.

Area:

India = 6.0 M ha

U.P. = 8.1 lakh ha

Electro-chemical Changes:

pH:

The depletion of oxygen, accumulation of CO_2 and an attach of reduction process specially effect the pH value of soil with excess moisture. The changes in soil pH not only affect the availability of plant nutrients but also affect the solubility of elements toxic to plant. Addition of excess moisture, an increase in pH of 1.60 units in 30 days may be observed in an acid laterite sandy loam soil. Reduction of ferric and manganic compounds and release of OH^-, production of ammonia and CO_2 account for pH rises in acid soils. In calcareous soils, excess moisture produces the reverse effect, namely a decrease in pH.

Redox potential: Redox potential is regulated by the concentration of reduced and oxidized substances according to the following equation:

$$Eh = E_0 + \frac{RT}{nF} \ln (Ox)/Red.$$

Where,

E_o = Standard redox potential

R = Gas constant

T = Absolute temperature

F = Faraday constant

n = Valency

ln = Natural logarithm

Ox = Concentration of oxidized substances

Red. = Concentration of reduced substances

The electrochemical property that differentiates a soil with excess moisture from an upland soils is its redox potential. The aerobic soils are characterized by high positive potential in the range of +500 to +700 mv. After addition of excess moisture, there is a sharp drop in the potential and within soil with excess moisture potentials are recorded in most soils. The soil with excess moisture may slow potentials as low as –300 to -400 mv depending upon the type of soil, organic matter, Fe, and Mn content, pH and temperature.

With excess moisture, the anaerobic organisms utilize the oxidized soil compounds as electron acceptors resulting into reduction of these compounds. Organic matter serves as an electron donar and accelerates the reduction process.

NO_3 and Mn oxides are reduced at high potential, while at extremely low potentials sulphate may also be reduced.

So these soils can be managed by adopting above-mentioned practice for sustainable agricultural production.

Redox potential of following:

	Redox potential (mv)
Good aerated soil	+ 400 to + 700
Oxidised surface soil	+ 300 to + 500
NO_3^- to N_2	+ 280 to + 220
Mn^{+++} to Mn^{++}	+ 280 to + 220
Fe^{+++} to Fe^{++}	+ 180 to + 150
O_2 to H_2O	+ 380 to + 320
Reduced zone	+ 200 to - 400
CO_2 to CH_4	- 200 to - 280
SO_4^{--} to S^{--}	- 120 to - 180
Waterlogged soil	- 250 to - 350

Tree species suitable for various problematic soils:

Salt affected soils	Acidic soils	Waterlogged soils	Marshy areas
Acacia auriculiformis	*Acacia nilotica*	*Acacia nilotica*	*Eucalyptus robusta*
Acacia nilotica	*Albizia procera*	*Albizia procera*	*E. salligna*
Albizia procera	*Albizza lebbeck*	*Anthocephalus indicus*	*E. rudis*
Casuarina equisetifolia	*Terminahia arjuna*	*Butea monosperma*	*Bringtonia spp.*
Emblica officinalis	*Prosopis julifora*	*Dalbergia sissoo*	
Pongamia pinnata	*Tectona grandis*	*Eucalyptus pinnata*	Coastal area
Prosopis cineraria	*Eucalyptus tereticornis*	*Salix tetrasperma*	*Pongamia glabra,*
Zizyphus nummularia		*Eucalyptus camaldulensis*	*Avecenia officinalis*
Albizia lebbeek		*Syzyagium cumini*	*Eucalyptus tereticornis*
Cassia fistula		Terminalia arjuna	
Prosopis juliflora			
Butea monosperma			

Salt affected soils
Melia azedarach
Salvadora oleoides
Acacia farnesiana
Azadirachta indica
Eucalyptus hybrid
Parkinsonia acculeata
Terminalia arjuna

CHAPTER-4

SOIL EROSION

Soil erosion has been our nation's most devastating environmental disaster. Soil in quantities of incomprehensible magnitude has been lost from the land where its forfeited productivity could have contributed to our national wealth. Soil erosion is the detachment and transportation of soil material from one place to another through the action of wind, water in motion or by the beating action of raindrops. Soil erosion is caused by detachment and removal of soil particles from land surface. With the advent of modem civilization, the pressure on land increased, which, leads to it's over exploitation, and subsequently, its degradation.

Types of Soil Erosion:

The process of soil erosion goes through two main stages of activities, namely the detachment of soil particles and their transportation. The principal agents responsible for these two stages of activities are wind, water and gravity, Wind erosion occurs in arid and semi-arid areas where rainfall is scanty and the day temperatures are very high. Water erosion occurs in areas where rainfall is high and gravity erosion in areas that are near to pits, rivers, roads, etc.

The types of soil erosion caused by three natural agents-winds, water and gravity are described below.

Erosion may be broadly classified into two groups:

(A) Geological erosion:

Geological erosion, sometimes referred to as natural or normal erosion, represent erosion under the cover of vegetation. It takes place as a result of the action of rainfall, runoff, topography, wind velocity, atmospheric temperature and gravitational forces.

- It is continuous and slow but constructive process.
- Soil removes is fairly balanced with soil formation.
- Erosion caused by biological activities is called as normal erosion.

(B) Accelerated erosion:

Accelerated erosion of soil is caused mainly by managerial errors, such as raising of crops without adopting any soil conservation practice, deforestation etc. It leads to erosion in excess of the threshold value of the new soil formation causing severe deterioration of the top surface of the land

- Its occur in faster rate and destructive process.
- Erosion in which rate of soil loss exceeds the rate of soil formation.

Water erosion:

Erosion of soil by water is caused by its two forms- liquid as the flowing water, and solid as the glaciers. The impact of rainfall causes splash erosion. Runoff water causes scarping and transport of soil particles leading to sheet, rill and gully erosion. Water waves cause erosion of bank sides, reservoirs, lakes and ocean. The sub-surface runoff soil erosion in the form of pipe erosion, which is termed as tunnel erosion. The glacial erosion causes heavy landslides. In India, glacial erosions are mainly confined to Himalayan regions.

- Movement of soil by rain water running rapidly over the exposed land surface.
- Most common found in humid and sub-humid areas where rainfall is excess and no vegetation over the land.
- In water erosion, greater soil fertility loss.

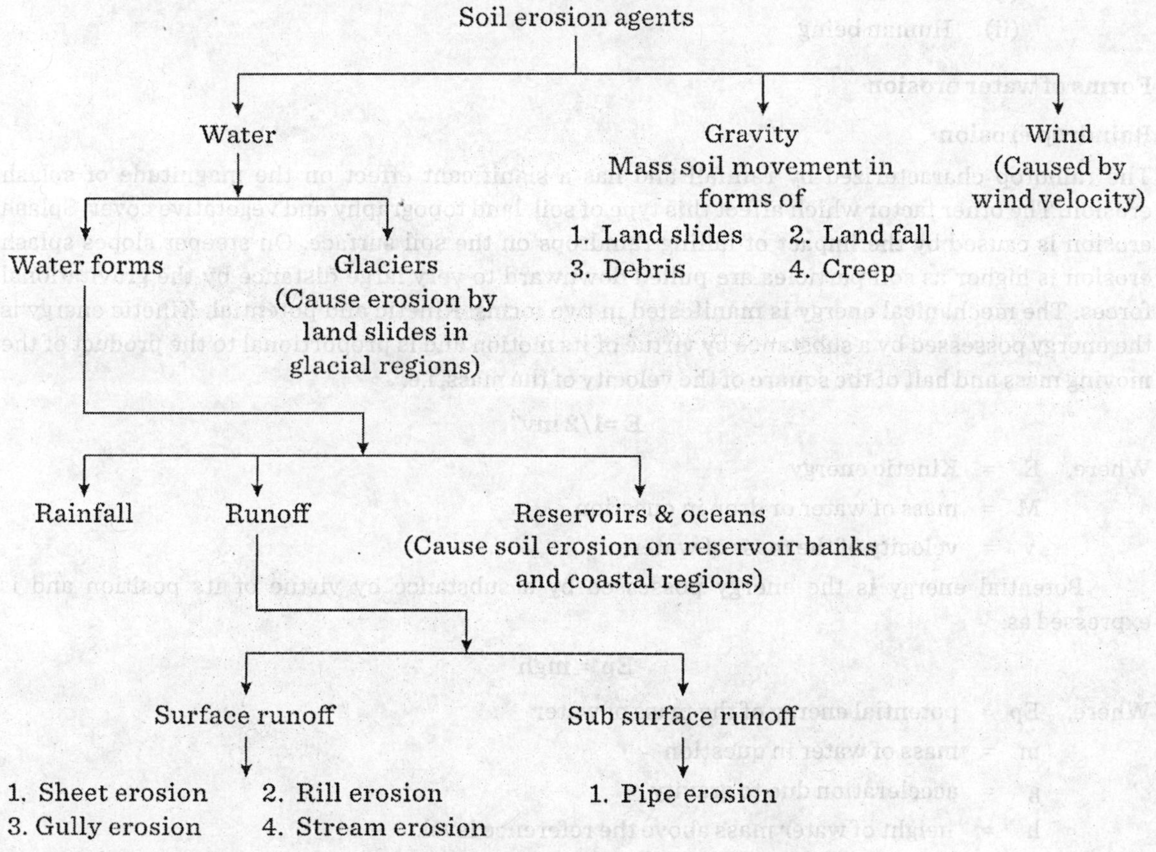

Causes of Soil Erosion:

The causes of soil erosion are as follows-

1. **Destruction of natural vegetation**

 (A) Deforestation

 (i) Shifting cultivation (ii) Forest fire

 (B) Over grazing

 (i) Excess grazing (ii) Non-rotational

 (C) Burning of grass land

2. Mismanagement of land

(A) Improper land management

(i) Up and down hill cultivation on slopy land

(ii) Un-successive growing of grass, crops

(iii) Removal of organic matter and plant nutrients

(iv) Faulty irrigation systems i.e. flood irrigation on slopy land

(v) Growing of wider row crops on slopy land

(B) Pressure on agriculture

(i) Animal

(ii) Human being

Forms of water erosion

Raindrop erosion

The raindrop characterized by rainfall and has a significant effect on the magnitude of splash erosion. The other factor which affect this type of soil, land topography and vegetative cover. Splash erosion is caused by the impact of falling raindrops on the soil surface. On steeper slopes splash erosion is higher as soil particles are pulled downward to very large distance by the gravitational forces. The mechanical energy is manifested in two forms: Kinetic and potential. Kinetic energy is the energy possessed by a substance by virtue of its motion and is proportional to the product of the moving mass and half of the square of the velocity of the mass, i.e.

$$E = 1/2\, mv^2$$

Where, E = Kinetic energy

M = mass of water or drop in question

v = velocity of the mass of water.

Potential energy is the energy possessed by a substance by virtue of its position and is expressed as:

$$Ep = mgh$$

Where, Ep = potential energy of the mass of water

m = mass of water in question

g = acceleration due to gravity

h = height of water mass above the reference level.

- Splash erosion is caused by the impact of falling raindrops.
- Splash erosion influenced by rainfall intensity and rainfall size.
- Splash erosion takes place, when rainfall intensity exceeds infiltration rate.

Sheet erosion:

This is caused by the unconfined flow of water running across the surface. The effects of sheet erosion are often hard to distinguish because such thin layers of soil are being removed.

- Removal of a thin and fairly uniform layer of surface soil by the action of rainfall and runoff water.
- Sheet erosion takes place on smooth and regular filled slope.

- Shallow soils suffer greater reduction in production than deep soil.
- Extremely harmful to the land
- Splash erosion and sheet erosion occur side by side.
- In India, about 60 lakh ha land affected by sheet erosion.

Rill erosion:

It is caused by water concentrating into innumerable, closely-spaced small channels. The removal of soil by water from small channels which develop due to concentration of surface flow along the slope. It generally develops on bare and unprotected land where soil conservation measures are not adopted. In advanced stage, it may take the shape of gullies. The rill erosion is an intermediary stage between sheet erosion and gully erosion.

- Removal of surface soil by running water with the formation of narrow shallow channels that can be leveled by normal cultivation
- Rill erosion is more serious in soils having a loose shallow top soil.

Gully erosion:

Gullies are steep-sided trenches formed by the coalescence of many rills. Once they started are difficult to stop. It is the advanced stage of rill erosion which cannot be obliterated by normal tillage operations. Rills with more than 30 cm depth are generally called gullies. Gullies obstruct the operation of farm machinery and tend to deepen and widen with every heavy rainfall. They cut up large fields into small fragments and in course of time, make them unfit for cultivation. The rate of gully erosion depends on several factors, viz. runoff producing characteristics of the watershed, drainage area: soil characteristics, channel alignment, slope and vegetative cover on the ground, etc.

The development of gully occurs due to following four stages:

1. **Formation stage:** Channel erosion takes place by downward scour of the topsoil. If the topsoil can provide resistance, this stage proceeds slowly.
2. **Development stage:** Upstream movement of the gully head and simultaneous enlargement of width and depth take place. The weak parent material is rapidly removed by cutting C horizon.
3. **Healing stage:** Vegetation begins to grow in the channel and further erosion ceases.
4. **Stabilization stage:** The gully bed and sides reach a stable slope and sufficient vegetation grows to anchor the soil and to trap soil flowing from upstream. New topsoil develops and vegetative cover grows.

Classification of Gullies:

(A) Based on the size of the gully:

	Very small gullies (g_1)	Small gullies (g_2)	Medium gullies (g_3)	Deep and narrow gullies (g_4)
Deep	up to 3 m	up to 3 m	3-9 m	>9 m
Bed width	Not > 18 m	> 18	Not < 18 m	Varies
Slide slope	vary	8-15 %	8-15%	Mostly steep

(B) Based on the shape of the gully:

Particulars	U shaped gullies	V shaped gullies
Location	Alluvial plains where surface and sub-surface soils are easily erodible	Hilly regions where land surface is tough to resist the cutting
Cross section	U shaped	V shaped
Slope	Slopy field	High slopy field
Velocity of flow	Less	High
Catchment's area	Large	Small
Runoff volume/ unit time	More	Less
Lateral spacing of gullies	Large	Small
Control measures	Permanent drop structure	Check dams

(C) Based on state of the gully:

Active gullies	Inactive gullies
Active gullies are those whose dimensions are enlarged with time	Inactive gullies are those whose dimensions are constant with time
These gullies are found in plain	These gullies are found in rocky areas.

Other forms of erosion:

Besides the four main water erosions described above, there are, some special forms of erosion also. These are discussed below:

Landslides or Slip erosion:

During monsoon season or heavy down-pour, the hills get saturated with water, resulting in instability of the land mass. The complete surface portion of soil may slips down from, its actual position causing landslides. Landslides are very common in hills and huge quantities of soil are washed away in streams and rivers. Hills devoid of vegetation are more prone to slip erosion, causing damage to agricultural lands, orchards, communication systems and human lives. Minor landslips occur during the drying phase in the months of May and June, especially in the sandy hills of Shiwaliks.

Stream bank erosion:

It is caused by streams and torrents, also called *chos,* cutting their banks and adjoining fields. It is mainly due to runoff flowing over the side of the stream banks or by scouring and undercutting below the water surface, the latter process being more serious. It is influenced by deforestation and overgrazing. Scour erosion is influenced by the velocity and direction of water flow, depth and width of channel and soil texture.

River bank erosion:

This is a common phenomenon in the rivers of the eastern and western Himalayan regions. During floods, the rivers undermine their banks and big chunks of land mass are engulfed by flood water. This leads to broadening of river beds and changes in their courses.

Coastal erosion:

The strong waves of sea strike against the seashore causing large scale soil erosion. In this case, erosion is a combined effect of wind and water.

Harmful effect of water erosion:

1. Loss of top fertile soils
2. Accumulation of sand or other unproductive coarse soil materials on other productive lands
3. Silting of lakes and reservoirs
4. Silting of drainage and irrigation channels
5. Decrease watertable
6. Fragmentation of land especially in gully erosion
7. Increase in flood (heavy floods in the river)
8. Adverse effect on public health
9. Loss of nutrients
10. Destroy soil structure
11. Adverse effect on crop yield
12. Hindrance of farming operations
13. Affect on transportation
14. Destruction of vegetation

Wind erosion:

Wind erosion is the detachment and transportation of soil particles by the forces of moving wind. In regions, where the precipitation is low and the atmospheric temperature during the day, high, the velocity of wind is invariably very high. Such climatic conditions are generally prevalent in arid and semi-arid regions, where the wind velocity is also very high.

Wind erosion is caused by mismanagement of land resources, such as intensive farming, over grazing, deforestation, etc.

i) Movement of soil by wind rapidly over the exposed land surface.
ii) Most common found in arid and semi-arid areas where rainfall is insufficient and no vegetation over the land.
iii) Less fertility loss.

Mechanism of wind erosion:

Transport of soil particles through wind by three processes namely; surface creep, saltation and suspension, depending on the size of soil particles that are moved by wind.

Surface creep:

The movement of larger size particles (> 0.5 mm) occurs by the process of surface creep, where the movement is by rolling and sliding, of soil particles on the land surface caused by direct push of wind and collisions between the saltating particles, i.e. the particles which are already moving by the process of saltation.

- Movement of big size particles (0.5-1.0 or 2 mm) by rolling or sliding along the land surface, activated by wind.

- Surface creep process destroys soil structure.

Saltation:

It is next step of movement taken by soil particles which are in process of surface creep, after these particles have gained sufficient speed. On gaining speed, these particles start to hop, leap, and bounce off the ground, and the movement by this process is called saltation.

- Soil particles of medium size (0.1-0.5 mm) are carried by wind in series of short jump/bounces/hop/ leap/ lift caused by direct pressure of wind on soil particles.
- Soil particles can jump up to the height of 1.0 m (average 30-60 cm).
- Saltation process is less in rough surface.
- Jump particle moves 200-1000 round/minute.

Suspension:

Very small-sized particles of soil (0.1 mm) are carried by suspension in the air. The soil particles stay in suspension, because the downward movements of soil particles due to their weight are balanced by the lifting forces of eddies in the fluid.

- When soil particles are very small (<0.1 mm) and they are carried over long distance.
- Floating of small sized particles in air.
- They are carried in suspension up to 100 km.
- The particles come down only with rains or when wind velocity is decreased.
- Makes smooth surface.

Forms of wind erosion:

There are five different forms of wind erosion are recognized

1. **Abration/ Corrasion:**
   i) Abration form is found in clody soils.
   ii) With the effect of strong winds, the clods are contact and broken up into small pieces.
2. **Efflation:** This form is found in silty and clay particles dominant in soil.
3. **Efflution:** This form of wind erosion is found in calcareous soil where clay and sand particles are dominant. With the effect of strong wind clay particles carry away a long distance by suspension process and big particles or sliding i.e. sand carry away a short distance by rolling or sliding.
4. **Detrusion:** This form is found where bunds are constructed in cultivated soil with the effect of strong wind, soil particles of bund slip down and deposited in lower place.
5. **Extrusion:** This form of wind erosion is found in sandy soils where sands particles are dominant. With the effect of strong wind, sand particles carry away from one place to another by rolling/ sliding.

Harmful effects of wind erosion:

1. Loss of soil and plant nutrients
2. Loss of production
3. Change of soil textural class
4. Reduction in water holding capacity of the soil
5. Exposure of hard rocks
6. Formation of sand dunes
7. Destroy soil structure

Factors affecting wind soil erosion:

The following factors affecting soil erosion by wind are assessed by the Wind Erosion Equation (WEE). These are given below:

- soil erodibility
- surface roughness
- climate
- length of field
- vegetative cover

1.	Tillage	:	Soils are loosened by the excess tillage	=	More erosion
2.	Structure	:	Single grain structured soil	=	More erosion
3.	Organic matter	:	Low organic matter in soil	=	More erosion
4.	Vegetation	:	Soil covered with vegetation	=	Reduce erosion
5.	Weather	:	Continuous dry weather	=	Increase erosion
6.	Wet and moist soils			=	decrease erosion
7.	Wind velocity	:	High wind velocity	=	Increase erosion
8.	Soil	:	Rough surface	=	Least wind erosion
9.	Smooth surface			=	More wind erosion
10.	Clody soil			=	Less wind erosion
11.	Without clods			=	More wind erosion

Factors affecting soil erosion:

Numerous factors affect soil erosion depending upon the local conditions with regard to physical, chemical and biological nature of the soil. The major factors affecting soil erosion are:

- Climate, especially rainfall and wind velocity and their intensity in time and space
- Topography, slope of land, its nature, degree and length
- Physical, chemical and biological characteristics of the soil
- Nature of ground cover and land use pattern
- Biotic interference due to human beings and animals
- Conservation practices, and
- Mass erosion, such as earthquakes, land slides, torrents, floods, etc.

Mathematically, soil erosion may be depicted by functional equation:

Erosion = f (cl, t, v,s,h,......)

Climate:

The major climatic factors influencing runoff and erosion are rainfall, temperature, and wind. Rainfall is by far the most important factor causing soil erosion by water. The soil erosion depends on the intensity, kinetic energy, amount, duration and frequency of rainfall. The intensity, duration and frequency of rainfall govern the rate and the volume of runoff. A light rain that can be easily absorbed in the soil causes no runoff and soil loss. When the intensity of rain is more than infiltration rate, it causes runoff and soil loss.

- The following are the factors in the soil erosion process:

- The detaching capacity of the erosive agent.
- The detachability of the soil.
- The transporting capacity of the erosive agent.
- The transportability of the soil.

Topography

On flat lands, erosion is usually not a problem. It is the sloping lands that experience increasingly greater problems of erosion. The degree of slope and the length of slope are the two main features of topography affecting erosion. Velocity of the runoff water is influenced mainly by the degree of slope. If the land slope is increased four times, the velocity of water flowing over it is approximately doubled, When the velocity is doubled the erosive capacity, as represented by the kinetic energy of the flowing water, is increased about four times. The land slope, its degree and length, is important in determining the extent of soil erosion. Higher the degree of slope and longer is its length, more, is the soil erosion due to increased velocity of water running down the slope.

Vegetation:

Vegetative cover retards the surface velocity of runoff water and transportation capacity reduces. Also the water remains on the surface for longer period and more amount infiltrates into the ground. The root system of the plants acts as binding agent for the soils and the detachability reduces. The decaying of roots, leaves etc. increases the porosity of soil and the capacity to absorb water increases. Addition of humus increases the biological activity inside the soil. Accelerated biological activity is favourable for growth of bacteria, earthworms etc. This also creates a favourable condition for absorption and holding of more moisture inside the soil. Thus vegetative cover helps to conserve both soil and moisture in several ways. The nature of vegetative cover on the soil surface is a major factor in influencing soil erosion. The following are the major effects of vegetation on soil erosion.

❖ **Interception of rainfall:**

A part of the rainfall intercepted by the canopy of vegetation never reaches the soil, but is evaporated directly from the leaves and stems. This part of the rainfall does not contribute to the runoff. In addition, the vegetative canopy absorbs the impact or the raindrops and thereby minimizes the dispersion of soil by the raindrops.

❖ **Runoff velocity:**

Vegetative cover is a hindrance to runoff. A well-distributed, close-growing vegetation not only slowdown the rate at which water travels down the slope but also tends to prevent a concentration of water. These two effects greatly reduce the erosive capacity of the runoff water.

❖ **Root effects:**

The knitting and binding effect of root systems in the surface layer of soils, aggregates the soil into granules and increases its resistance to erosion.

❖ **Biological influence:**

The soil fauna are most active in soils having ample vegetative cover. The soil under a thick forest covers increase aeration and good environment for the activities of beneficial bacteria and earthworms.

Soil characteristics:

Erodibility of soil is influenced by its physical and chemical properties, including soil texture, structure, and organic matter, nature of clay and amounts kinds of salts present. Soil structure, texture, organic matter infiltration and permeability influence the runoff and soil loss. Fine soils are more susceptible to erosion than coarse soils. The organic matter helps in binding the soil particles and improves the water holding capacity of soil. The three most significant soil characteristics which influence soil erosion are: (1) infiltration capacity, (ii) structural stability, and (iii) antecedent soil moisture.

Biological factors:

Biological factors that influence the soil erosion are the activities of man and animal. Felling of trees, faulty cultivation practices like up and down cultivation in hilly areas, overgrazing by cattle etc. contribute much to the soil erosion. Without such activities, the soil, climate, vegetation etc. will remain in a balanced condition and no erosion will take place.

Control of wind erosion:

Any practice or measure that reduces the wind velocity or improves the soil characteristics is helpful to control wind erosion. Improved soil characteristics should have better structure, improved cohesive property, and good moisture holding capacity. Some of the measures may provide both the requirements. Vegetation improves the soil structure and at the same time retards the surface wind velocity. In general the following practices may be followed to control the wind erosion.

i. The soil should be covered with vegetation or crop residues as far as possible.
ii. Limited cultivation should be done.
iii. Dry soils should not be tilled.
iv. Permanent vegetation may be established on unproductive soils.
v. After the rains, the soil may be tilled so that clod formation takes place.
vi. Tillage implements should be selected in such a manner that rough surface is formed and crop residue is not buried.
vii. Overgrazing should be avoided.

Principal methods of reducing surface wind velocity are vegetative control, tillage practices and mechanical method. Vegetative control consists of cultivated crops, field and strip cropping; stubble mulching, shrubs and tress although from part of the vegetation, act as mechanical barrier to wind. Other mechanical barriers are windbreaks also may be used, some important practices are discussed below:

Strip cropping:

Field and contour strip cropping consists of alternate strips of row (erosion-susceptible) crops and close growing (erosion-resistant) crops in the same field. The strip cropping is laid out generally parallel to the field boundary or perpendicular to the erosive wind direction. The main benefits of strip cropping are:

- Vegetation provides physical protection against blowing of soil.
- Soil erosion is limited to a distance equal to the width of the erosion susceptible crop.
- Better conservation of moisture takes place.
- Particles carried in saltation are trapped.

Main problems in strip cropping are:

- In a mechanized farm, movement of machinery becomes difficult due to narrow strips.
- In case of attack by insects, there is more number of edges for protection.

The width of the strips should be selected in a way such that the farming operation is not hampered and at the same time much erosion does not take place. For example, in a sandy soil the width of the erosion susceptible crop should be limited to 6 m. But for movement of machinery, the width has to be increased. In a sandy loam soil the width can be increased up to 30 m.

Windbreak:

A windbreak is defined as type of barrier for protection from winds and refers to any mechanical or vegetative barriers of buildings, gardens, orchards and feed lots. Windbreaks made up of just mechanical barriers are not very useful for field crops. However, they are frequently used for the protection of farm sheds and small areas. The mechanical barriers include brush fences, board walls, vertical burlap or paper strips. Brush matting, rock or gravel barriers are also used as windbreaks. Some of these barriers are impermeable and others are semi-impermeable. Generally the semi-impermeable barriers are more useful as they provide better diffusion and eddying effects on the leeward side of the barrier. When vegetable crops in organic soils are required to be protected, vertical burlap or paper strips are often used. Brush matting, debris, rock, gravel etc. are more useful for stabilizing sand dune areas.

Shelterbelts (Shrubs and Trees):

A shelterbelt usually consisting of shrubs and trees is a longer barrier than windbreak. It is primarily used for protection of field crops, soils and conservation of moisture. The shelterbelt is not only useful for wind erosion control, but also like; windbreak saves fuel, increases livestock production, reduces evaporation, prevents firing of crops from hot winds. In addition, it may provide better fruiting in orchards, make spraying of trees for insect control more effective manner.

Tillage practices:

Tillage practices if properly adopted can reduce the soil blowing to a great extent, similarly, faulty tillage operations increase the soil erosion by wind. If the soil is pulverized and the crop residues are buried due tillage operations, erosion problem increases. The effective way of prevention of wind erosion is by producing a rough, cloddy surface and exposing the crop residues on the surface. If the land is ploughed at optimum moisture content after the rains, big clod and large aggregate formation takes place, which ultimately decreases the process of erosion.

Predicting Soil Erosion:

Definition: By taking all factors into account, a prediction equation was developed for calculating the soil loss, called as **Universal Soil Loss Equation** (USLE):

- It accounts all parameters affecting the soils loss.
- It predicts the annual soil loss.
- It computes sheet erosion.
- It is not used when slope are steeper than 20%.

A = RKLSCP *By Wischmeier & Smith* (1965)

Where, A = Estimated gross erosion (t/ha/year)

R = Rainfall erosivity factor
L = Soil erodibility (For calculation of water erosion)
S = Slope gradient factor
C = Crop cover or crop management factor
P = Supporting conservation practice factor

Demerits of USLE:

1. This equation does not compute the sediment yield from the watershed, directly.
2. Annual sediment yield can be determined which is less important for design of water storage structures.
3. This equation is used in small watershed to prediction of soil loss.

Modified Universal soil loss equation (MUSLE):

i) This equation computes sediment yield either monthly or seasonal.

ii) This equation is used in large watershed for prediction of soil loss.

$$Y = 95\,(Q.q_p)^{0.56}$$ *KLSCP By Williams* (1975)

Where, Y = Sediment yield for an individual storm (t/ha)
Q = Volume of runoff (ha-m)
qp = Peak flow rate (m^3)

E = f (IKCLV) *By Chepil and Woodruffridowy* (1965)

Where, E = Average annual soil loss (t/ha/year)
I = Soil erodibility
K = Roughness
C = Control of soil conservation measure
L = Slope length
V = Vegetation

Note: Soil erosion = f (Erosivity of rain) × (Erodibilty of soil)

CHAPTER-5

WATERSHED MANAGEMENT

Watershed: Definition. In physical terms, a watershed is a hydrological entity and refers to an area above a given drainage point. The Total area contributing water flowing to the outlet is the watershed at that outlet. Watershed attributes an dudes size, shape, orientation, topography, geology, climate, surface condition, land use, ground water and Socio-economic aspects.

There are various definitions of watershed have been cited by different investigators/ agencies, few of them are mentioned as under:

1. It is a divide, which separates one watershed from the other.
2. The watershed is that area, which is drained by a stream flow originated in the area is drained through a single outlet.
3. The drainage basin is the area, which contributes runoff to the main stream and its tributaries.
4. It is a hydrological entity, in which hydrologic cycle complete.
5. As per oxford dictionary, the watershed is the separating ridge between two river systems.
6. The watershed and drainage basin are synonymous term indicating an area surrounded by a ridge line that is drained through a single outlet. When it refers to the area that contributes the runoff for a river it a stream, them it is referred as catchment of the river.
7. Chow stated that the watershed is a divide separating one drainage basin from another. Drainage basin is referred as a catchment in British literature.

Watershed Management:

Definition: Watershed management implies, the judicious use of all the resources i.e., land, vegetation and water of the watershed to achieve maximum production with minimum hazard to the natural resources and for the well being of the people. The management should be carried out on the watershed basis.

Scope: Soil erosion and land degradation in India is a major problem in India. Out of total geographical area of 328 m ha 175 m ha (53% of total) are categorized degraded land of which 150 m ha are subjected to serious water and wind erosion, furthermore, 106 m ha are considered to be dry lands.

Objectives:

(i) To control damaging runoff and degradation and thereby conservation of soil and water.

(ii) To manage and utilize the runoff for useful purposes of watershed development.

(iii) To protect, conserve and improve the land of watershed for more efficient and sustainable production.

(iv) To protect and enhance the water resources originating in the watersheds.

(v) To check the soil erosion and reduce the effect of sediment yield on the watershed.
(vi) To rehabilitate the deteriorating lands
(vii) To moderate the flood peaks at downstream areas
(viii) To establish watershed management practices and measures
(ix) To enhance the ground water recharge, wherever applicable.
(x) To improve and increase the production to timbers, ranges and wild life resources.
(xi) To expand electricity extension activities.
(xii) To increase the intensity of agricultural extension activities.

Needs of Watershed Management

(i) Increasing agricultural Production.
(ii) Controlling of flood.
(iii) Reducing sediment load.
(iv) Increasing forest production.

Factors Affecting Watershed Management

(a) **Watershed characteristics**
- (i) Shape and size
- (ii) Topography
- (iii) Relief
- (iv) Soils

(b) **Climatic characteristics**
- (i) Precipitation
- (ii) Amount and intensity of rainfall

(c) **Watershed operation**

(d) **Land use pattern**
- (i) Vegetative cover
- (ii) Density
- (iii) State i.e. type and quality

(e) **Social status of inhabitants**

(f) **Water resources and their capabilities.**

Watershed Planning

Basic information:

(i) Statistics of production and live stock
(ii) Pattern of land ownership
(iii) Topography, land use cropping systems and yields and potential capability of land .
(iv) Data on vegetation, rainfall, runoff, erosion problem and round water.
(v) Information on exhisting water sources like ponds tanks & wells etc.
(vi) Servies faciiliities such as banking schools, market input supply, extension agencies, health and veterinary facilities.

Watershed problems:

(a) **Flood Damage:**
- (i) Amount and value of land improvement and other properties exposed to the flood hazards in the watershed

(ii) Frequency of flood occurrence

(iii) Significance of small frequent floods or large in frequents floods in total flood problem.

(iv) Limitation.

(v) Other pertinent problem.

(b) Sediment damage:

(i) Problem of reservoir sedimentation.

(ii) Problem of channel silting

(iii) Drainage problem

(iv) Irrigation development

(v) Loss of agricultural land

(c) Erosion damage:

(i) Extent of sheet erosion, gully and channel erosion.

(ii) Down stream damage due to sediment deposition.

(ii) Effect on agricultural production due to erosion.

(iii) General effects on Watershed's economy.

(d) Water Management Problem:

(i) Irrigation needs

(ii) Drainage

(iii) Water supply for agricultural and non-agricultural uses.

(iv) Other management needs.

(e) Special Problems:

(i) Land slip

(ii) Land Slide

(iii) Torrents

(iv) Highway erosion

(v) Mines

Management Plan:

(1) Management problem for Agricultural Lands:

(a) Agronomic measures:

(i) Crops proposed

(ii) Crop rotation

(iii) Cultural operations

(iv) Introduction of improved seeds

(v) Use of manure and fertilizers.

(vi) Plant protection measures

(vii) Expected yield obtained

(viii) Expected cost invested.

(b) Engineering Measures:

(i) Proposed measures and their justification.

(ii) Specification of individual measures with plant and design

(iii) Cost estimation.

(2) Management programme for conservation Irrigation:

(i) Selection and design of irrigation method for each part of land

(ii) Irrigation duration

(iii) Conveyance system

(iv) Location and smoothing operation.

(v) Plant for drainage.

(vi) Agronomical measures.

(3) Management programme for drainage:

(i) Types of drainage system.

(ii) Specification

(iii) Location of outlets and their design

(iv) Intercepting and relief drains along with their design.

(v) Maintenance.

(vi) Cost-estimation

(4) Management programme, for flood protection:

(i) Channel improvement

(ii) Flood retarding structures- their design and cost estimation.

(5) Management problems of forest lands:

(i) Management practices for existing forests, degraded forests scrubs and farm forestry.

(ii) Engineering measures for forest and range lands.

(6) Management programmes for Grazing :

(a) Pastures:

(i) Selection of grass species. (ii) Raising methods.

(iii) Mechanical practices viz. (iv) Yields

(v) Cost invested

(b) Grass lands:

(i) Use of improve practices (ii) Grazing system

(7) Programme for special problem:

(a) Land slide: retaining wall on other structures.

(b) Stream Bank erosion-vegetative or structural methods.

(c) Erosion along the highway- vegetation or revetment.

(d) Gullies and ravines- vegetation and gully control structures.

(e) Social status of inhabitants:-

(f) Water resources and their capabilities

3. Factors affecting watershed management:

(a) Watershed Characteristics

(i) Shape and size (ii) Topography Relief

(iii) Soils

(b) Climatic characteristic:

(i) Precipitation (ii) Amount and intensity of rainfall

(c) Watershed operation

(d) Land use pattern:-

(i) Vegetative cover (ii) Density

(iii) State is type and quality

Soil Conservation measures foe watershed development:

(a) Vegetative measures (agronomical practices) :

(i) Strip cropping
(ii) Pasture farming
(iii) Fertilizing the crop land
(iv) Grass land farming etc.

(b) Engineering measures

(i) Gully Plugging
(ii) Diversions
(iii) Water ponds
(iv) Reservoirs
(v) Drainage work
(vi) Flood protection
(vii) Ground water recharging structures
(viii) Terracing & bunding
(ix) Land levelling and grading
(x) Grassed water ways

Principal and practices of dry land Agricultural:

The concept of watershed development in dry land has a special significance. The development programme involves many of the watershed attributes (listed in definition) and also includes component such as special crop management (inter-cropping and bed system etc) soil and water conservation, water harvesting (for lifesaving irrigation) alternate land use etc. clearly, these factors are inter-related and often have an effect on the social environment as well. The major functions of the practices adopted of dry land agriculture are:

(i) Control of runoff
(ii) Reduction of raindrop impact
(iii) Improvement in soil structure.

River valley projects: River valley projects serves:

(i) Irrigation and water supply
(ii) Flood control and navigation
(iii) Power generation
(iv) Others (production of wild life recreation, drainage etc.)

Silting of Reservoirs and Dams:

(i) Slope of stream
(ii) Reservoir length
(iii) Reservoir construction
(iv) Size of sediment
(v) Capacity inflow
(vi) Reservoir operation
(vii) Inflow patterns of the stream
(viii) Vegetal growth
(ix) Sediment load in river flow
(x) Shape of reservoirs
(xi) Outlets
(xii) Sediment-reservoirs volume ratio.

Reservoir sediment control:

(a) Pre-constructive measures:

(i) Selection of the dam site
(ii) Design of reservoir
(iii) Construction of check dam
(iv) Installation of vegetative screen
(v) Construction of sluice gate under the dam
(vi) Reservoir operation
(vii) Erosion control

(b) Post-constructive measures:

(i) Removal of flood water (ii) Stirring of sediment

(iii) Removal of silt deposits

Flood Control:

(a) Reducing flood flows

(i) Watershed treatment (ii) Reservoirs treatment

(b) Increasing channel capacity

(i) Channel improvement (ii) Increasing cross section

(iii) Increasing velocity

Planning of Watershed Works:

(A) Watershed Description:

The description of watershed should be so explained, that can represent a clear picture of the watershed's condition. As for as possible, sufficient information's on physical features of watershed and problem encountered, should be fully included to descried the watershed. Mainly the following contents are considered for at watershed description:

Location: To describe the location of watershed, the information's such as: name of river basin tributary, physiographic region, principal communication lines associated latitude and longitude of watershed are collected.

Size and shape: The size of watershed either in the unit of sq. km or ha. and shape of watershed either it is long and narrow or fan shape, whatever those are, should be mentioned.

Climate: About the climate of watershed the following details are usually collected:

1. Precipitation: forms, annual, seasonal and monthly distribution, storm pattern, rainfall intensity, duration and its area; distribution.
2. Temperature: maximum and minimum, soil temperature etc.
3. Evaporation
4. Relative humidity
5. Wind velocity and its duration
6. Solar radiation.

Geology : The geology characteristics of watershed is also considered as an important factor to describe the watershed undertaken for management .It includes the geology information such as: nature of parent rocks, fractures, faults, weathering , ground water charge and extent of outcrops.

Slope: It includes the degree of slope length of slope mean slope of the land surface and proportion of different areas of watershed falling under various slope groups.

Surface drainage: Regarding drainage features of watershed, the following information are required to collect for describing the physical features of watershed.

1. Nature of stream flow (i.e. whether it is perinneal, intermittent, ephemeral, spring fed or seasonal)
2. **Drainage network:** Stream orders, density and length of streams etc.
3. Morphological characteristics of streams.

Soils: The information about major soils groups existing in the watershed and their hydrologic grouping, physical and chemical properties of the soils etc. are collected for needful.

Physiography: It consists of noting the details of elevation of different parts of watershed, mountainous range etc. for describing the watershed.

Watershed needs: Under this head, the contents such as sources of surface and sub-surface (ground water) water, water use for domestic, irrigation, power generation and recreation, future needs of water exploration etc. are recorded.

Land use and cover conditions:

It includes following details, used to describe the watershed:

(a) Existing land use cover condition:

(i) Forest lands (ii) Range lands (iii) Cultivated lands
(iv) Waste lands (v) Habitations (vi) Miscellaneous uses.

(b) Forests:

(i) Types and area under each type of forest
(ii) Hydrologic condition (iii) Legal status
(iv) Present management (v) Area under regeneration
(vi) Filling and logging practices (vii) Rights and forest fires

(c) Range lands:

(i) Extent and their major classification
(ii) Closure (iii) Grazing practices
(iv) Grazing incidence (v) Cattle population (number and types)
(vi) Important grass species their distribution and hydrological

(d) Agricultural Lands:

(i) Extent (ii) Land use capability classification
(iii) Area under each class and sub-class (iv) Major crops
(v) Rain fed areas with suitable crops grown (vi) Orchards and their area

Economics data: The economics and social data are needed to determine the cost-benefit ratio of the project formulated for watershed management under this, the following data are collected:

(i) Economic condition of the population
(ii) Professions and dependence on resources
(iii) Market and marketing practices
(iv) Return from various resources e.g. from forests, range lands, agricultural lands including irrigation as well as rain fed etc.

Watershed problems:

The problems such as : flood drought erosion and sediment damage and other problems related to the conservation, development, utilization, disposal water originating in the major problems are considered as under,

Flood damage: The following points are considered to determine or evaluate the flood damage occurred in watershed:

(1.) Amount and value of lands improvements and other properties exposed to the flood hazards in the watershed.

(2) Frequency of flood occurrence.

(3) Significance of small frequent floods or large infrequent floods in total flood problem.

(4) Limitations.

(5) Other pertinent problem.

Sediment damage: The problem of exposed be sediment deposition is considered in following cases:

1. Problem of reservoir sedimentation
2. Problem of channel silting drainage problem
3. Irrigation development
4. Loss of agricultural land.

Erosion damage: The problem erosion damage is studied under the following content,

1. Extent of sheet erosion, gully and channel erosion.
2. Downstream damage due to sediment deposition
3. Effects on agricultural production due to erosion.
4. General effects on watershed's economy.

Watershed management Problem: It includes the details on irrigation needs, drainage, water supply required for agricultural and non-agricultural uses and other management needs.

Special problem: The problem such as: lands slip, land slide, torrents, highway erosion, mine etc. are counted for preparation of watershed works plan.

(B) Proposed Management Programmers:

Under this section, the management programmes to be used for fighting against watershed problems encountered, are described. The different proposed management programmes are listed as under:

1. Management programme for Agricultural lands:

To control deterioration of agricultural land, the various control measures for them are divided into following two groups:

(a) **Agronomic measures**— these are as under :

(i) Crop proposed	(ii) Crops rotation
(iii) Cultural operations	(iv) Introduction of improved seeds
(v) Use of manures and fertilize	(vi) Plant protection measures
(vii) Expected yield obtained	(viii) Expected cost invested

(b) **Engineering measures:**

(i) Proposed measures and their justifications.

(ii) Specification of individual measures with plan and design

(iii) Cost estimation

2. **Management Programmers of Conservation Irrigation:**
   (i) Selection and design of irrigation methods for each part of land.
   (ii) Irrigation duration
   (iii) Conveyance system
   (iv) Location and specification of system
   (v) Smoothening operation
   (vi) Plan for drainage

Agro forestry System on Degraded Lands:

Dry	Moist
1. Silvipastoral in place of natural savanna on degraded forest lands	1. Energy farms for high economic output of biomass to meet fodder and firewood needs
2. Horti-silvipastoral system on marginal agriculture lands	2. Silvipastoral wiith species diversity and intensive management
3. Agri-silcipastoral use of rainfed farming areas	3. Agri-silvipastoral with perennial crop species
4. Multi-tree-stylo combinations to derive nursing effect from legume and short rotation tree species	4. Tree-crop mixture for higher biomass production and nutrient turnover
5. Farm boundary planting with crop lands to get improved microenvironment	5. Fruit trees on farm boundary mixed with fast growing MPTS
6. MPTS mixtures around homesteads for shade, fruit feed	6. MPTS around homesteads for fruit like Artocarpus heterophylla
7. Suitable species *Acacia albida, A farnesiana A tortills, Albizia armara Azadirachta indiaca borassus flabellifer Dalbergia sissoo Hardwichia Melia azedarch Pongamia pinnata Prosopis cinerarea Prosopis juliflora*	7. *Acacia auriculiformis. A. decurrens A mangium, A nilotica ,albizia falcataria,a lebbeck calliaqndra calothyrsus areca catchu, calliaandra calothuyrsus Casuarina oligodon Cocos nucifera Gliricdia sepium, Gmelina aborea Leucaena leucocephala Melia azedarach Morus alba Populus deltoids, Sesbania* spp.

Agro forestry and Land Improvement

Agro forestry refers to land-use systems in which trees or shrubs are grown in association with agricultural crops, pastures or livestock, and in which there are both ecological and economic interactions between trees and other components. It is the economical that are the most distinctive feature, and which differentiate Agro forestry from social forestry (forestry carried out by communities or individuals), although there is a large overlap.

It is emerging as an appropriate form of land use in recent years. It has special significance in the Asia-Pacific region because of following reasons:

(i) Deterioration in land-man ratio- containing 69% of world's population but only 28% of world's agricultural land consequently per capita available of agricultural land in the region is only 0.27 ha as compared to 1.64 ha in rest of world .

(ii) Around 2 million ha of forests are deforested annual in the region and about 75 % of this deforested area is used for agriculture- resulting in many cases in spontaneous unplanned agriculture. With destruction of forest and depletion of the mantle of fertile top soils .these soils are quickly rendered unproductive.

(iii) Agroforestry allows farmers to diversity their production system, thus reducing the risk of total crop failure and increase their income potential.

The willingness of a farmer to adopt Agro forestry will depend on the importance he places on the likely costs and benefits .Agro forestry has numerous potential benefits for rural farmers including:

- Improved soils, though the prevention of erosion ,nutrient cycling, the addition of organic matter and nitrogen fixation
- Improved animal production through better quality fodder, fodder provision during dry periods and climatic change
- Cash income from tree products such as fruit and timer
- Reduced risk through diversification of products
- The availability of fuel wood (and hence reduced tine or cost of collecting fuel)
- Timber for building and fence poles
- Demarcation of boundaries and live fences using trees
- Protection from the wind
- A more scenic landscape.
- Agro forestry also has certain costs associated with it, such as;
- The time required for planting and managing trees.
- The space occupied by the trees
- Competition between trees and other crops for light, nutrient and water may result in reduced crops.
- Possible insect and animal pests associated with trees or a changed microclimate.

In most Agro forestry systems tree products are used to enhance the animal or crop enterprise (with the exception of tree products such as coffee or tea that are grown as a cash crop). For example the flow of tree products to other enterprises mat be in the form of mulch and fodder.

Agroforestry practices have potential not only to conserve the production base of the land but also tend to improve it.

Rates of erosion in tropical forests and tree crop systems:

Land use	Erosion (t/ha/yr)		
	Minimum	Medium	Maximum
Multistory tree garden	0.01	0.06	0.14
Natural rain forests	0.03	0.30	6.16
Shifting cultivation (fallow period)	0.05	0.15	7.40
Forest plantation (undisturbed)	0.02	0.58	6.20
Tree crops with cover crop or mulch	0.10	0.75	5.60
Shifting cultivation (cropping period)	0.40	2.78	70.05

Land use	Erosion (t/ha/yr)		
	Minimum	Medium	Maximum
Taungya cultivation period	0.63	5.23	17.37
Tree crops clean weeded	1.20	47.60	182.90
Forest plantations, burned litter removed	5.92	53.40	104.80

Patterns of runoff and soil from different production systems at Jhansi:

(Average of 5 years: 1989-92)

Production system	% Runoff (rainfall)	Soil loss (t/ha)
Bare plot	38.08	23.92
Natural grassland	13.68	4.36
Improved grassland	12.68	3.01
Sown pasture	8.86	1.99

Potential agro forestry practices for control of soil erosion:

Practices	Environment in which Potentially Applicable
1. Plantation crop combination	Humid to moist sub-humid climate
2. Multistory tree gardens (and home gardens)	Humid and moist sub-humid climate. May be possible in drier regions also.
3. Hedgerow intercrossing (and barrier hedges)	Humid and sub-humid climates. May be possible in semi-arid climate also.
4. Trees on erosion control structures	Any
5. Windbreaks and shelterbelts	Semi-arid regions
6. Silvipastoral systems	Arid, semi-arid and sub-humid regions some humid regions also (especially S. America)
7. Reclamation forestry leading to multiple use	Any
8. Combination of above in integrated watershed management	Any

Potential agro forestry practices in relations to soil fertility:

1. **Practices with substantial positive effect on soil fertility**
   - improved tree fallow
   - trees on croplands
   - plantation crop combinations
   - homegardens
   - hedgerow intercrossing
   - trees on erosion control structures
   - windbreak and shelterbelts
   - biomass transfer
   - trees on gangland or pastures

- woodlots with multipurpose management
- reclamation forestry leading to multiple use

2. Practices with small positive of neutral effect on soil fertility

- boundary planting
- plantation crop with pastures

3. Practices with positive or negative effect on soil fertility

- shifting cultivation

4. Practices with neutral or negative effect on soil fertility

- taungya

CHAPTER-6

VARIOUS ISO TERM

- **Asogon line:** It is an imaginary connecting points of same specified angle or same wind direçtion.
- **Asotherm:** It is an imaginary line connecting points of equal temperature.
- **Contour line:** It is an imaginary line representing points of equal elevation (height) above a given level, such as mean sea level.
- **dsometric:** It is an imaginary line connecting points of equal variations of pressure.
- **Isentropic:** It is an imaginary line connecting points of equal entropy
- **Isobar:** It is an imaginary line connecting points of equal atmospheric pressure.
- **Isobath:** It is an imaginary line on a map connecting points of equal groundwater depth.
- **Isobathytherm:** It is an imaginary line connecting points of equal temperature and depth underground
- **Isobront:** It is an imaginary line connecting points of simultaneous storm development
- **Isochar line:** It is an imaginary connecting points of similar distinguishing characteristics of plant life
- **Isocheim:** It is an imaginary line connecting points of same average winter temperature
- **Isochor:** It is an imaginary line connecting points of equal volume.
- **Isochrone:** It is an imaginary line connecting points of equal travelling time from a point.
- **Isodose:** It is an imaginary line connecting points that receive equal doses of radiation
- **Isodrosotherm:** It is an imaginary line connecting points of equal dew point
- **isogram**. It is an imaginary line representing points of equal value.
- **Isohaline:** It is an imaginary line drawn on a map or chart to indicate connecting points of equal salinity in the ocean.
- **Isohel:** It is an imaginary line on a map connecting points having the same duration of sunshine.
- **Isohyet:** It is an imaginary line on a map connecting points having the same amount of rainfall in a given period.
- **Isoline:** it is also called as contour line or isarithm or
- **Isoneph:** It is an imaginary line representing points of equal amounts of cloud cover.
- **Isopach:** It is an imaginary line connecting points of equal thickness of geological strata
- **Isophene:** It is an imaginary line connecting points of the same phenotype or seasonal variation.

- **Isopleth:** It is an imaginary line on a map connecting points having equal depth of evapotranspiration
- **Isoseismal:** It is an imaginary line connecting points of same earthquake intensity.
- **Isosine:** It is an imaginary line representing points of equal slope
- **Isotachs:** It is an imaginary line representing points of equal wind speed.
- **Isothere:** It is an imaginary line connecting points of equal summer temperature.
- **Isotherombrose:** It is an imaginary line connecting points of equal ratio of summer rainfall to annual rainfall.
- **Isocline:** It is a curve through points at which the parent function's slope will always be the same, regardless of initial conditions.
- **Soerodent :** It is an imaginary line connecting points areas of equal erosion potentials.

CHAPTER-7

DEFINITIONS

- **Abutment:** The sloping sides of a valley that supports the ends of a dam.
- **Accelerated erosion:** Erosion much more rapid than normal or geologic erosion, primarily as a result of the activities of man.
- **Apron:** A pad of non-erosive material designed to prevent scour holes developing at the outlet ends of culverts, outlet pipes, grade stabilization structures, and other water control devices.
- **Aquifer:** An underground porous, water-bearing geological formation. The term is generally restricted to materials capable of yielding an appreciable supply of water.
- **ASTM:** American Society for Testing Materials, an association that publishes standards and requirements for materials used in the construction industry.
- **Backwater:** The rise in water surface elevation caused by some obstruction such as a narrow bridge opening, buildings or fill material that limits the area through which the water shall flow.
- **Barrel:** A conduit placed through a dam, levee, or dike to control the release of water.
- **Base Flood Elevation (BFE):** The water surface elevation corresponding to a flood having a one percent probability of being equalled or exceeded in a given year.
- **Base Flow:** Stream discharge derived from groundwater sources as differentiated from surface runoff. Sometimes considered to include flows from regulated lakes or reservoirs.
- **Bearing Capacity:** The maximum load that a material can support before failing.
- **Benchmark:** A marked point of known elevation from which other elevations may be established.
- **Berm:** A narrow shelf or flat area that breaks the contiguity of a slope.
- **Best Management Practices:** Design, construction, and maintenance practices and criteria for Storm water facilities that minimize the impact of storm water runoff rates and volumes, prevent erosion, and capture pollutants.
- **Centerline of Channel:** The middle point or baseline of a channel.
- **Channel erosion:** An erosion process whereby the volume and velocity of flow wears away the bed and/or banks of a well-defined channel.
- **Channel Improvement:** Alteration, maintenance, or reconstruction of the channel area for the purpose of improving the channel capacity or overall drainage efficiency. The noted "improvement" does not necessarily imply water quality or habitat improvement within the channel or its adjacent area.
- **Channel Stabilization:** Protecting the sides and bed of a channel from erosion by controlling flow velocities and flow directions using jetties, drops, or other structures and/or by fining the channel with vegetation, riprap, concrete, or other suitable lining material.

- **Channel:** A portion of a natural or artificial watercourse which periodically or continuously contains moving water, or which forms a connecting link between two bodies of water. It has a defined bed and banks which serve to confine the water.
- **Channelization:** Alteration of a stream channel by widening, deepening, straightening, or paving certain areas to improve flow characteristics.
- **Chute:** A high-velocity, open channel (usually paved) for conveying water down a steep slope without erosion.
- **Cohesion:** The capacity of a soil to resist shearing stress, exclusive of functional resistance.
- **Contour Line:** Line on a map which represents a contour or points of equal elevation.
- **Contour:** An imaginary line on the surface of the earth connecting points of the same elevation.
- **Control Structure:** A structure designed to control the rate of flow that passes through the structure, given a specific upstream and downstream water surface elevation.
- **Cross-Section:** A graph or plot of ground elevation across a stream valley or a portion of it, usually along a line perpendicular to the stream or direction of flow.
- **Culvert:** A closed conduit used for the conveyance of surface drainage water under a roadway, railroad, canal or other impediment.
- **Cut:** (1) A portion of land surface or area from which earth has been removed or will be removed by excavating. (2) The depth below the original ground surface to the excavated surface.
- **Cut-and-Fill:** The process of earth grading by excavating part of a higher area and using the excavated material for fill to raise the surface of an adjacent lower area.
- **Cutoff Trench:** A long, narrow excavation (keyway) constructed along the center line of a dam, dike, levee, or embankment and filled with relatively impervious material intended to reduce seepage of water through porous strata.
- **Cutting:** A detached leaf stem or piece of root that is encouraged to form roots. A greenwood cutting is made during the period of active growth. A hardwood cutting is made during the dormant season.
- **Dam:** A barrier to confine or impound water for storage or diversion, to prevent gully erosion, or to retain soil, sediment, or other debris.
- **Datum:** Any level surface to which elevations are referred, usually using Mean Sea Level.
- Design Life: The period of time for which a facility is expected to perform its intended function.
- **Design Storm:** A selected storm event, described in terms of the probability of occurring once within a given number of years, for which drainage or flood control improvements are designed and built.
- **Detention Storage:** The temporary detaining of storage of storm water in storage facilities, on rooftops, in streets, parking lots, school yards, parks, open spaces or other areas under predetermined and controlled conditions, with the rate of release regulated by appropriately installed devices.
- **Detention Time:** The theoretical time required to displace the contents of a tank or unit at a given rate of discharge (volume divided by rate of discharge).
- **Detention:** Managing storm water runoff by temporary holding and controlled release.

- **Dibble Bar:** A heavy metal tool with a blade and foot pedal used to open holes for planting seeds, sprigs, cuttings or seedlings.
- **Dike:** An embankment to confine or control water. Often built along the banks of a river to prevent overflow of lowlands; a levee.
- **Discharge:** Usually the rate of water flow. A volume of fluid passing a point per unit time commonly expressed as cubic feet per second, cubic meters per second, gallons per minute, or millions of gallons per day.
- **Ditch:** A man-made, open drainage way in or into which excess surface water or groundwater drained from land, storm water runoff, or floodwaters flow either continuously or intermittently.
- **Diversion:** A channel with a supporting ridge on the lower side constructed at the top, across, or bottom of a slope for the purpose of controlling surface runoff.
- **Drain.** A buried slotted or perforated pipe or other conduit (subsurface drain) or a ditch (open drain) for carrying off surplus groundwater or surface water.
- **Drainage (soil):** As a natural condition of the soil, drainage refers to both the frequency and duration of periods when the soil is free of saturation. Soil drainage conditions are defined as:
- **Drainage Area:** The area draining into a stream at a given point. It may be of different sizes for surface runoff, subsurface flow and base flow, but generally the surface runoff area is considered as the drainage area.
- **Drainage Board:** A board consisting of three to five persons including the county executive (commissioners) or members appointed by the executive body (at least one of the Board member must be a county executive). The County Surveyor serves on the Board as an ex-officio, non- voting member. In a county having a consolidated city, the department of public works of the consolidated city comprises the drainage board. The Board is responsible for adopting drain classifications and a long-range plan, and for making decisions regarding the design, construction, reconstruction, and/or maintenance of regulated drains in the county.
- **Drainage Improvement:** An activity within or adjacent to a natural stream or a man-made drain primarily intended to improve the flow capacity, drainage, erosion and sedimentation control, or stability of the drainage way.
- **Drainage way:** A natural or artificial stream, closed conduit, or depression that carries surface water. This term has been used throughout the Handbook as a neutral term applying to all types of drains and watercourses, whether man-made or natural.
- **Drainage:** The removal of excess surface water or groundwater from land by means of ditches or subsurface drains. Also see Natural drainage.
- **Drawdown:** Lowering of the water surface in an open channel or lake or groundwater.
- **Drop Inlet:** A structure in which water drops through a vertical riser connected to a discharge conduit or storm sewer.
- **Drop Spillway:** A structure in which the water drops over a vertical wall onto an apron at a lower elevation.
- **Drop Structure:** A structure for dropping water to a lower level and dissipating its surplus energy without erosion.
- **Duration:** The time period of a rainfall event.
- **Earth Dam:** A dam constructed of compacted suitable soil materials.

- **Earth Embankment:** A man-made deposit of soil, rock, or other material often used to form an impoundment:
- **Emergency Spillway:** Usually a vegetated earth channel used to safely convey flood discharges around an impoundment structure.
- **Energy Dissipater:** A device used to reduce the energy of flowing water to prevent erosion.
- **Environment:** The sum total of all the external conditions that may act upon a living organism or community to influence its development or existence.
- **Erodibility:** Susceptibility to erosion.
- **Erosion:** The wearing away of the land surface by water, wind, ice, gravity, or other geological agents. The following terms are used to describe different types of water erosion:
- **Excess Rainfall:** The amount of rainfall that runs directly off an area.
- **Farm or Field Tile:** A small diameter clay pipe installed in an agricultural area to allow drainage of farmland.
- **Flood Frequency:** A statistical expression of the average time period between floods equalling or exceeding a given magnitude. For example, a 100-year flood has a magnitude expected to be equalled or exceeded on the average of once every hundred years; such a flood has a one- percent chance of being equalled or exceeded in any given year. Often used interchangeably with "recurrence interval".
- **Flood or Flood Waters:** A general and temporary condition of partial or complete inundation of normally dry land areas from the overflow, the unusual and rapid accumulation, or the runoff of surface waters from any source.
- **Flood Peak:** The highest stage or greatest discharge attained by a flood event, thus peak stage or peak discharge.
- **Flood plain:** The channel proper and the areas adjoining the channel which have been or hereafter may be covered by the regulatory or 100-year flood. Any normally dry land area that is susceptible to being inundated by water from any natural source. The floodplain includes both the floodway and the floodway fringe districts.
- **Flood Stage:** The stage at which overflow of the natural banks of a stream begins.
- **Floodway:** The channel of a river or stream and those portions of the flood plains adjoining the channel which are reasonably required to efficiently carry and discharge the peak flow of the regulatory flood of any river or stream.
- **Flume:** A constructed channel lined with erosion-resistant materials used to convey water on steep grades without erosion.
- **Foundation Drain:** A pipe or series of pipes that collects groundwater from the foundation or footing of structures to improve stability.
- **Freeboard:** A vertical distance between the elevation of the design high-water and the top of a dam, diversion ridge, or other water control device.
- **French Drain:** A drainage trench backfilled with a coarse, water-transmitting material may contain a perforated pipe.
- **Gabion:** A wire mesh cage, usually rectangular, filled with rock and used to protect channel banks and other sloping areas from erosion.

- **Gauge:**
  - A device for measuring precipitation, water level, discharge, velocity, pressure, temperature, etc.
  - A measure of the thickness of metal.
- **Gauging Station:** A selected section of a stream channel equipped with a gauge, stage recorder, or other facilities for determining stream stage and discharge.
- **Gradation:** The distribution of the various sized particles that constitute sediment, soil, or other material, such as riprap.
- **Grassed Waterway:** A natural or constructed waterway, usually broad and shallow, covered with erosion-resistant grasses and used to safely conduct surface water from an area.
- **Gully erosion:** An erosion process whereby runoff water accumulates in narrow channels and, over relatively short periods, removes the soil to considerable depths, ranging from 1-2 ft. to as much as 75-100 ft.
- **Head Loss:** Energy loss due to friction, eddies, changes in velocity, elevation, or direction of flow.
- **Head water:** (1) The source of a stream. (2) The water upstream from a structure or point on a stream.
- **Head:** The height of water above any plane of reference.
- **Hydrograph:** A graph showing for a given point on a stream the discharge, stage (depth), velocity, or other property of water with respect to time.
- **Hydrologic Cycle:** The circuit of water movement from atmosphere to earth back to the atmosphere through various stages or processes, such as precipitation, runoff, infiltration, percolation, storage, evaporation, and transpiration.
- **Hydrology:** The science of the behaviour of water in the atmosphere, on the surface of the earth, and underground. A typical hydrologic study is undertaken to compute flow rates associated with specified flood events.
- **Hydromulching:** The process of applying mulch hydraulically in a water medium.
- **Hydroseeder:** The machine/equipment used to disseminate seed hydraulically in a water medium. Mulch, lime, and fertilizer can be incorporated into the sprayed mixture.
- **Impoundment:** Generally, an artificial water storage area, such as a reservoir, pit, dugout, sump, etc.
- **Keyway:** A cut-off trench dug beneath the entire length of a dam to cut through soil layers that may cause seepage and possible dam failure.
- **Lag Time:** The interval between the center of mass of the storm precipitation and peak flow of the resultant run-off.
- **Laminar Flow:** Flow at relatively slow velocity in which fluid particles slide smoothly along straight lines everywhere parallel to the axis of a channel or pipe.
- **Land Capability:** The suitability of land for use. Land capability classification involves consideration of: (1) the risks of damage from erosion and other causes and. (2) the difficulties in land use owing to physical land characteristics, including climate.
- **Land Surveyor:** A person licensed under the laws of the State of Indiana to practice land surveying.

- **Land Use Controls:** Methods for regulating the uses to which a given land area maybe put, including such things as zoning, subdivision regulation, and floodplain regulation.
- **Loam:** A soil textural classification in which the proportions of sand, silt, and clay are well balanced. Loams have the best properties for cultivation of plants.
- **Mean Depth:** (1) Average depth. (2) Cross-sectional area of a stream or channel divided by its surface or top width.
- **Mean Velocity:** Average velocity of a stream flowing in a channel or conduit at a given cross-section or in a given reach. It is equal to the discharge divided by the cross-sectional area of the reach.
- **Moderately well drained:** Water is removed from the soil somewhat slowly resulting in small but significant periods of wetness, and mottling occurs between 18 and 36 in.
- **Mulch:** A natural or artificial layer of plant residue or other materials covering the land surface which conserves moisture, holds soil in place, aids in establishing plant cover, and minimizes temperature fluctuations.
- **Mutual Drain:** A drain that: (1) Is located on two or more tracts of land that are under different ownership; (2) Was established by the mutual consent of all the owners; and (3) Was not established under or made subject to any drainage statute.
- **Natural Drainage:** The flow patterns of storm water run-off over the land in its pre-development state.
- **Normal Depth:** Depth of flow in an open conduit during uniform flow for the given conditions.
- **Open Drain:** A natural watercourse or constructed open channel that conveys drainage water.
- **Outfall:** The point, location, or structure where wastewater or drainage discharges from a pipe or open drain to a receiving body of water.
- **Outlet Channel:** A waterway constructed or altered primarily to carry water from man-made structures, such as smaller channels, tile lines, and diversions.
- **Outlet:** The point of water disposal from a stream, river, lake, tidewater, or artificial drain.
- **Outside Valley:** The spacing or width of corrugations for corrugated metal pipe.
- **Overland Flow:** Consists of sheet flow, shallow concentrated flow and open channel flow.
- **Peak Discharge:** The maximum instantaneous flow from a given storm condition at a specific location.
- **Percolation Rate:** The rate, usually expressed as inches per hour or inches per day, at which water moves through the soil profile.
- **Perennial Stream:** A stream that maintains water in its channel throughout the year.
- **Permeability (soil):** The quality of a soil that enables water or air to move through it. Usually expressed in 'inches per hour or inches per day.
- **Permeability Rate:** The rate at which water will move through a saturated soil. Permeability rates are classified as:
    - Very slow— Less than 0.06 in/hr.
    - Slow— 0.06 to 0.20 in/hr.
    - Moderately slow— 0.20 to 0.63 in/hr.
    - Moderate— 0.63 to 2.0 in/hr.

- Moderately rapid— 2.0 to 6.3 in/hr.
- Rapid— 6.3 to 20.0 in./hr.
- Very rapid— More than 20.0 in/hr.

- **Poorly drained:** Water is removed so slowly that it is wet for a large part of the time, and mottling occurs between 0 and 8 in.
- **Reach:** The smallest subdivision of the drainage system, consisting of a uniform length of open channel. Also, a discrete portion of river, stream or creek. For modelling purposes, a reach is somewhat homogeneous in its physical characteristics.
- **Receiving Stream:** The body of water into which runoff or effluent is discharged.
- **Recharge Basin:** A basin provided to increase infiltration for the purpose of replenishing groundwater supplies.
- Recharge. Replenishment of groundwater reservoirs by infiltration and transmission from the outcrop of an aquifer or from permeable soils.
- **Recurrence Interval:** A statistical expression of the average time between floods equalling or exceeding a given magnitude.
- **Regulated Drain:** A drain, either open channel or closed tile/sewer, subject to the provisions of the Indiana Drainage Code, I.C.-36-9-27.
- **Regulatory Flood:** The discharge or elevation associated with the 100-year flood as calculated by a method and procedure which is acceptable to and approved by the Indiana Department of Natural Resources and the Federal Emergency Management Agency. The "regulatory flood" is also known as the "base flood".
- **Reservoir:** A natural or artificially created pond, lake or other space used for storage, regulation or control of water. May be either permanent or temporary. The term is also used in the hydrologic modelling of storage facilities.
- **Retention Facility:** A facility designed to completely retain a specified amount of storm water runoff without release except by means of evaporation, infiltration or pumping. The volumes are often referred to in units of acre-feet.
- **Retention:** The storage of storm water to prevent it from leaving the development site. May be temporary or permanent.
- **Revetment:** Facing of stone or other material, either permanent or temporary placed along the edge of a stream to stabilize the bank and protect it from the erosive action of the stream. Also see Revetment riprap.
- **Rill erosion:** An erosion process in which numerous small channels only several inches deep are formed; occurs mainly on recently disturbed and exposed soils (see Rill).
- **Rill:** A small intermittent watercourse with steep sides, usually only a few inches deep.
- **Riparian Rights:** A principle of common law requiring that any user of waters adjoining or flowing through his lands must use and protect them in a manner that will enable his neighbour to utilize the same waters undiminished in quantity and undefiled in quality.
- **Riprap:** Broken rock, cobble, or boulders placed on earth surfaces, such as the face of a dam or the bank of a stream, for protection against the action of water (waves). Revetment riprap is material graded such that: (1) no individual piece weighs more than 120 lbs. and (2) 90-100% will pass through a 12-inch sieve, 20-60% through a 6-inch sieve, and not more than 10% through a 1½-inch sieve.

- **Riser:** The inlet portions of a drop inlet spillway that extend vertically from the pipe conduit barrel to the water surface.
- **River Restoration:** Restoring the channel of a stream or ditch to its perceived original, non-obstructed capacity by means of clearing & snagging, obstruction removal, and inexpensive stream bank protection measures. The term "restoration", as noted, does not necessarily imply restoration or improvement of water quality or habitat within the channel or its adjacent area.
- **Riverine:** Relating to, formed by, or resembling a stream (including creeks and rivers).
- **Runoff:** That portion of precipitation that flows from a drainage area on the land surface, in open channels, or in storm water conveyance systems.
- **Sand:** (1) Soil particles between 0.05 and 2.0 mm in diameter. (2) A soil textural class inclusive of all soils that are at least 70% sand and 15% or less clay.
- **Saturation:** In soils, the point at which a soil or aquifer will no longer absorb any amount of water without losing an equal amount.
- **Scouring:** The clearing and digging action of flowing water, especially the downward erosion caused by stream water in seeping away mud and silt from the stream bed and outside bank of a curved channel.
- **Sediment Delivery Ratio:** The fraction of the soil eroded from upland sources that actually reach a stream channel or storage reservoir.
- **Sediment Discharge:** The quantity of sediment, measured in dry weight or by volume, transported through a stream cross-section in a given time. Sediment discharge consists of both suspended load and bed load.
- **Sediment Pool:** The reservoir space allotted to the accumulation of sediment during the life of the structure.
- **Sediment:** Solid material (both mineral and organic) that is in suspension, is being transported, or has been moved from its site of origin by air, water, gravity, or ice and has come to rest on the earth's surface.
- **Sedimentation:** The process that deposits soils, debris and other materials either on the ground surfaces or in bodies of water or watercourses.
- **Seedbed:** Soil prepared by natural or artificial means to promote the germination of seed and the growth of seedlings.
- **Seepage:** The passage of water or other fluid through a porous medium, such as the passage of water through an earth embankment.
- **Settling Basin:** An enlargement in the channel of a stream to permit the settling of debris carried in suspension.
- **Sheet erosion:** The gradual removal of a fairly uniform layer of soil from the land surface by runoff water.
- **Soil and Water Conservation District (SWCD):** A public organization created under state law as a special-purpose district to develop and carry out a program of soil, water, and related resource conservation, use, and development within its boundaries. A subdivision of state government with a local governing body.
- **Soil Horizon:** A horizontal layer of soil that, through processes of soil formation, has developed characteristics distinct from the layers above and below.

- **Soil Profile:** A vertical section of the soil from the surface through all horizons.
- **Soil Structure:** The relation of particles that impact to the whole soil a characteristic manner of breaking- e.g., crumb, block, platy, or columnar structure.
- **Soil Texture:** The physical structure or character of soil determined by the relative proportions of the soil separates (sand, silt, and clay) of which it is composed.
- **Soil:** The unconsolidated mineral and organic material on the immediate surface of the earth that serves as a natural medium for the growth of land plants. Also see alluvial soil, Clay, Cohesive soil, Loam, Permeability (soil), Sand, Silt, Soil horizon, Soil profile, Subsoil, Surface soil, Topsoil.
- **Somewhat poorly drained:** Water is removed from the soil slowly enough to keep it wet for significant periods but not all of the time, and mottling occurs between 8 to 18 in.
- **Specific Gravity:** The ratio of (1) the weight in air of a given volume of soil solids at a stated temperature to (2) the weight in air of an equal volume of distilled water at a stated temperature.
- **Spillway:** (1) A passage, such as a paved apron or channel, for surplus water over, around, or through a dam or similar structure. (2) An open or closed channel, or both, used to convey excess water from a reservoir. It may contain gates, either manually or automatically controlled. To regulate the discharge of excess water. Also see Emergency spillway, Principal spillway.
- **Splash erosion:** The spattering of small soil particles caused by the impact of raindrops on wet soils; the loosened and spattered particles may or may not be subsequently removed by surface runoff.
- **Storm Duration:** The length of time that water may be stored in any storm water control facility, computed from the time water first begins to be stored.
- **Storm Event:** An estimate of the expected amount of precipitation within a given period of time. For example, a 10-yr. frequency, 24-hr. duration storm event is a storm that has a 10% probability of occurring in any one year. Precipitation is measured over a 24-hr. period.
- **Storm Frequency:** The time interval between major storms of predetermined intensity and volumes of runoff— e.g., a 5-yr., 10-yr. or 20-yr. storm.
- **Storm Sewer:** A sewer that carries storm water, surface drainage, street wash, and other wash waters but excludes sewage and industrial wastes. Also called a storm drain.
- **Storm water Runoff:** The water derived from rains falling within a tributary basin, flowing over the surface of the ground or collected in channels or conduits.
- **Stream banks:** The usual boundaries (not the flood boundaries) of a stream channel. Right and left banks are named facing downstream.
- **Stream Gauging:** The quantitative determination of stream flow using gauges, current meters, weirs, or other measuring instruments at selected locations (see Gauging station').
- **Stream Length:** The length of a stream or ditch, expressed in miles, from the confluence of the stream or ditch with the receiving stream to the upstream extremity of the stream or ditch, as indicated by the solid or dashed, blue or purple line depicting the stream or ditch on the most current edition of the seven and one-half (7½) minute topographic quadrangle map published by the United States Geological Survey, measured along the meanders of the stream or ditch as depicted on the map.

- **Sub watershed:** A watershed subdivision of unspecified size that forms a convenient natural unit. See also Subarea.
- **Subarea/Sub basin:** Portion of a watershed divided into homogenous drainage units which can be modelled for purposes of determining runoff rates. The subareas/sub basins have distinct boundaries, as defined by the topography of the area.
- **Subsoil:** The B horizons of soils with distinct profiles. In soils with weak profile development, the subsoil can be defined as the soil below which roots do not normally grow.
- **Subsurface Drain:** A pervious backfield trench, usually containing stone and perforated pipe, for intercepting groundwater or seepage.
- **Surface Runoff:** Precipitation that flows onto the surfaces of roofs, streets, the ground, etc., and is not absorbed or retained by that surface but collects and runs off.
- **Tail water:** The water surface elevation at the downstream side of a hydraulic structure (i.e. culvert, bridge, weir, dam, etc.).
- **Tile Drain:** Pipe made of perforated plastic, burned clay, concrete, or similar material, laid to a designed grade and depth, to collect and carry excess water from the soil.
- **Tile Drainage:** Land drainage by means of a series of tile lines laid at a specified depth, grade, and spacing.
- **Time of Concentration (tc):** Is the travel time of a particle of water from the most hydraulically remote point in the contributing area to the point under study. This can be considered the sum of an overland flow time and times of travel in street gutters, , storm sewers, drainage channels, and all other drainage ways.
- **Toe of Dam:** The base or bottom of the sloping faces of a constructed dam at the point of intersection with the natural ground surface--normally a much flatter slope. A dam has an inside toe (the impoundment or upstream side) and an outside toe (the downstream side).
- **Toe of Slope:** The base or bottom of a slope at the point where the ground surface abruptly changes to a significantly flatter grade.
- **Topographic Map:** Graphical portrayal of the topographic features of a land area, showing both the horizontal distances between the features and their elevations above a given datum.
- **Topography:** The representation of a portion of the earth's surface showing natural and manmade features of a give locality such as rivers, streams, ditches, lakes, roads, buildings and most importantly, variations in ground elevations for the terrain of the area.
- **Unified Soil Classification System (USCS):** A system of classifying soils that is based on their identification according to particle size, gradation, plasticity index, and liquid limit.
- **Uniform Flow:** A state of steady flow when the mean velocity and cross-sectional area remain constant in all sections of a reach.
- **Vegetative Stabilization:** Protection of erodible or sediment producing areas with: permanent seeding (producing long-term vegetative cover), short-term seeding (producing temporary vegetative cover), or sodding (producing areas covered with a turf of perennial sod-forming grass)
- **Very poorly drained:** Water is removed so slowly that the water table remains at or near the surface for the greater part of the time; there may also be periods of surface ponding, the soil has a black to gray surface layer with mottles up to the surface.

- **Water Quality:** A term used to describe the chemical, physical, and biological characteristics of water, usually in respect to its suitability for a particular purpose.
- **Water Resources:** The supply of groundwater and surface water in a given area.
- **Water Table:** (1) The free surface of the groundwater. (2) That surface subject to atmospheric pressure under the ground, generally rising and failing with the season or from other conditions such as water withdrawal.
- **Watercourse:** Any river, stream, creek, brook, branch, natural or man-made drainage way in or into which storm water runoff or floodwaters flow either continuously or intermittently.
- **Watershed Area:** All land and water within the confines of a drainage divide. See also Watershed.
- **Watershed:** The region drained by or contributing water to a specific point that could be along a stream, lake or other storm water facilities. Watersheds are often broken down into subareas for the purpose of hydrologic modelling.
- **Weep Holes (engineering):** Openings left in retaining walls, aprons, linings, or foundations to permit drainage and reduce pressure.
- **Well-drained:** Excess water drains away rapidly, and no mottling occurs within 36 in. of the surface.

ABBREVIATIONS

AFM	:	Asymptotic Fit Method
AGNPS	:	Agricultural Non-Point Source Model
AIC	:	Akaike Information Criterion
AICc	:	Corrected Akaike's Information Criterion
AMC	:	Antecedent Moisture Condition
APEX	:	Agricultural Policy/Environmental Extender
ASCE	:	American Society of Civil Engineers Evapotranspiration
CN	:	Curve Number
DEM	:	Digital Elevation Model
EPA	:	Environmental Protection Agency
EPIC	:	Erosion-Productivity Impact Calculator
ERDAS	:	Earth Resources Data Analysis System
ET	:	Evapotranspiration
ETM	:	Enhanced Thematic Mapper
FAO	:	Food and Agricultural Organization
GDS	:	Gauge Discharge Site
GIS	:	Geographic Information System
GLEAMS	:	Groundwater Loading Effects of Agricultural Management Systems
HEC	:	Hydrologic Engineering Centre
HSG	:	Hydrologic Soil Group
ICAR	:	Indian Council of Agricultural Research
ILWIS	:	Integrated Land and Water Information System
IR	:	Infiltration Rate
IUSS	:	International Union of Soil Sciences
LER	:	Logarithm of Evidence Ratio
LULC	:	Land Use/Land Cover
MAE	:	Mean Absolute Error
MBE	:	Mean Bias Error
MNTTS	:	Minimum Temperature Time Series
MOM	:	Method of Moments
MXTTS	:	Maximum Temperature T time Series

NBSS & LUP	:	National Bureau of Soil Survey and Land Use Planning
NLEAP	:	Nitrate Leaching and Economic Analysis Package
NRCS	:	Natural Resources Conservation Service Curve Number
NRCS	:	USDA-Natural Resources Conservation Service
PCDS	:	Physical Catchment Descriptors
PET	:	Potential Evapotranspiration
RMSE	:	Root Mean Square Error
SCS-CN	:	Soil Conservation Service Curve Number
SMI	:	Soil-Moisture Index
SWAT	:	Soil and Water Assessment Tool
SWDC	:	State Water Data Centre
SWMM	:	Storm Water Management Model
USDA	:	United States Department of Agriculture
USLE	:	Universal Soil Loss Equation

Agricultural Research Institutes in India

- Central Agricultural Research Institute (CARI), Port Blair
- Central Arid Zone Research Institute, Jodhpur, Rajasthan
- Central Avian Research Institute, Izatnagar, Uttar Pradesh
- Central Inland Fishries Research Institute, Barrackpore, West Bengal
- Central Institute for Arid Horticulture, Bikaner, Rajasthan
- Central Institute for Cotton Research, Nagpur, Maharashtra
- Central Institute for Fisheries Education, Mumbai, Maharashtra
- Central Institute for Research on Buffaloes, Hissar, Haryana
- Central Institute for Research on Cotton Technology, Mumbai, Maharashtra
- Central Institute for Research on Goats, Farah, Uttar Pradesh
- Central Institute of Agricultural Engineering, Bhopal, Madhya Pradesh
- Central Institute of Brakishwater Aquaculture, Chennai, Tamilnadu
- Central Institute of Freshwater Aquaculture, Bhubaneswar, Orissa
- Central Institute of Post Harvest Engineering & Technology, Ludhiana, Punjab
- Central Marine Fisheries Research Institute, Kochin, Kerala
- Central Plantation Crops Research Institute, Kasaragod, Kerala
- Central Potato Research Institute, Shimla, Himachal Pradesh
- Central Research Institute for Dryland Agriculture, Hyderabad, Andhra Pradesh
- Central Research Institute for Jute and Allied Fiber, Barrackpore, West Bengal
- Central Rice Research Institute, Cuttack, Orissa
- Central Soil Salinity Research Institute, Karnal, Haryana
- Central Tuber Crops Research Institute, Thiruvananthapuram, Kerala
- ICAR Research Complex For Goa, Ela Old, Goa
- Indian Agricultural Research Institute, New Delhi
- Indian Agricultural Statistics Research Institute New Delhi
- Indian Grassland and Fodder Research Institute, Gwalior, Jhansi
- Indian Institute of Horticultural Research, Bangalore, Karnataka
- Indian Institute of Spices Research, Calicut, Kerala
- Indian Institute of Sugarcane Research, Lucknow, Uttar Pradesh
- Indian Institute of Vegetable Research, Varanasi, Uttar Pradesh

- Indian Institute of Soil Science , Bhopal, Madhya Pradesh
- Indian Lac Research Institute, Namkum, Ranchi
- National Academy of Agricultural Research Management, Hyderabad, Andhra Pradesh
- National Institute for Research on Jute & Allied Fibres Technology, West Bengal (formerly JTRL)
- National Institute of Animal Nutrition and Physiology, Bangalore, Karnataka
- Sugarcane Breeding Institute, Coimbatore, Tamilnadu
- Vivekanand Parvatiya Krishi Anusandhan Sansthan, Almora, Uttaranchal

Bureaus

- National Bureau of Agriculturally Important Microorganisms, Distt. Mau, Uttar Pradesh
- National Bureau of Animal Genetic Resources, Karnal, Haryana
- National Bureau of Fish Genetic Resources, Lucknow, Uttar Pradesh
- National Bureau of Plant Genetic Resources, New Delhi
- National Bureau of Soil Survey & Land Use Planning, Nagpur, Maharashtra

National Research Centres

- National Centre for Agril Economics & Policy Research, New Delhi
- National Research Centre for Agroforestry, Jhansi, Uttar Pradesh
- National Research Centre for Cachew, Puttur, Karnataka
- National Research Centre for Coldwater Fisheries, Bhimtal, Uttaranchal
- National Research Centre for Grapes, Pune, Maharashtra
- National Research Centre for Integrated Pest Management, New Delhi
- National Research Centre for Medicinal & Aromatic Plants, Boriavi, Gujarat
- National Research Centre for Mushroom, Solan, Himachal Pradesh
- National Research Centre for Onion and Garlic, Rajgurunagar, Pune, Maharashtra
- National Research Centre for Orchid, Pakyong, Sikkim
- National Research Centre for Sorghum, Hyderabad
- National Research Centre for Soybean, Madhya Pradesh
- National Research Centre for Weed Science, Madhya Pradesh
- National Research Centre for Yak, Dirang, Arunachal Pradesh
- National Research Centre on Camel, Bikaner, Rajasthan
- National Research Centre on Equines, Hissar, Haryana
- National Research Centre on Seed Spices, Ajmer

AGRICULTURAL UNIVERSITIES IN INDIA

State Agricultural Universities/Central University

1. Acharya NG Ranga Agricultural Univ., Rajendranagar, Hyderabad-500030, A.P.
2. Anand Agricultural University, Anand-388110, Gujarat
3. Assam Agricultural University, Jorhat-785013, Assam
4. Bidhan Chandra Krishi Viswavidyalaya, P.O Krishi Viswavidyalaya, Mohanpur, Nadia-741252, West Bengal
5. Birsa Agricultural University, Kanke, Ranchi- 834006, Jharkhand
6. Central Agricultural University, Imphal -795004, Manipur
7. Chandra Shekar Azad Univ. of Agriculture & Technology, Kanpur- 208002, U.P
8. Chaudhary Charan Singh Haryana Agricultural University, Hissar-125004,Haryana
9. Chaudhary Sarwan Kumar Himachal Pradesh Krishi Vishvavidhalaya, Palampur, Kangra- 176062, Himachal Pradesh
10. Dr Balasaheb Sawant Konkan Krishi Vidyapeeth, Dapoli, Ratnagiri-415712, Maharashtra
11. Dr Panjabrao Deshmukh Krishi Vidyapeeth, Krishi Nagar, Akola-444104, Maharashtra
12. Dr Yashwant Singh Parmar Univ. of Horticulture & Forestry, Solan, Nauni – 173230, Himachal Pradesh
13. Govind Ballabh Pant University of Agriculture & Technology, Pantnagar, Udhamsingh Nagar-263145, Uttarakhand
14. Guru Angad Dev University of Veterinary and Animal Sciences, Ludhiana-141004, Punjab
15. Indira Gandhi Krishi Vishwavidyalaya, Krishak Nagar, Raipur-492006, Chhattisgarh
16. Jawaharlal Nehru Krishi Vishwavidyalaya, Krishi Nagar, Jabalpur- 482004, M.P.
17. Junagadh Agriculture University, Moti Baug, Agril. Campus, Junagadh-362001, Gujarat
18. Karnataka Veterinary Animal and Fisheries Science University, P.B. No. 6, Nandinagar, Bidar-585401, Karnataka
19. Kerala Agricultural University, P.O Vellanikkara, Thrissur-680656, Kerala
20. Maharana Pratap Univ. of Agriculture & Technology, Udaipur-313001, Rajasthan
21. Maharashtra Animal Science & Fishery University, Nagpur, Maharashtra
22. Mahatma Phule Krishi Vidyapeeth, Rahuri-413722, Maharashtra
23. Marathwada Agricultural University, Parbhani -431402, Maharashtra
24. Narendra Deva University of Agriculture & Technology, Kumarganj, Faizabad-224229, Uttar Pradesh
25. Navsari Agricultural University, Vijalpore, Navsari-396450, Gujarat

26. Orissa Univ. of Agriculture & Technology, Siripur, Bhubaneswar-751003, Orissa
27. Punjab Agricultural University, Ludhiana -141004, Punjab
28. Rajasthan AgriculturalUniversity, Bikaner -334006, Rajasthan
29. Rajendra Agricultural University, Pusa, Samastipur-848125, Bihar
30. Sardar Vallabh Bhai Patel Univ of Agriculture & Technology, Modipuram, Meerut-250110, Uttar Pradesh
31. Sardarkrushinagar-Dantiwada Agricultural University, Sardarkrushinagar, Dantiwada, Banaskantha-385506, Gujarat
32. Sher-E-Kashmir Univ. of Agricultural Sciences & Technology, Railway Road, Jammu- 180012 (J&K)
33. Sher-E-Kashmir Univ. of Agricultural Sciences & Technology, Shalimar, Srinagar-191121, (J&K)
34. Sri Venkateswara Veterinary University, Tirupati, Chittoor- 517502, A.P.
35. Tamil Nadu Agricultural University, Coimbatore-641003, Tamil Nadu
36. Tamil Nadu Veterinary & Animal Sciences University, Madhavaram Milk Colony, Chennai- 600051, Tamil Nadu
37. University of Agricultural Sciences, Dharwad, Karnataka
38. University of Agricultural Sciences, Banglore- 560065, Karnataka
39. UP Pandit Deen Dayal Upadhaya Pashu Chikitsa Vigyan Vishwa Vidhyalaya Evam Go Anusandhan Sansthan, Mathura- 281001, Uttar Pradesh
40. Uttar Banga Krishi Vishwaviddyalaya, P.O. Pundibari, Distt. Cooch Behar-736165, West Bengal
41. Uttar Banga Krishi Vishwaviddyalaya, P.O. Pundibari, Distt. Cooch Behar-736165, West Bengal
42. West Bengal University of Animal & Fishery Sciences, 68 KB Sarani, Kolkata-700037, West Bengal
43. University of Horticultural Sciences, Venkataramnagudem, West Godavari, A.P.
44. University of Agricultural Sciences, Raichur-584102, Karnataka

Deemed-to-be Universities

1. Indian Agricultural Research Institute, Pusa-110012, New Delhi
2. Indian Veterinary Research Institute, Izatnagar, Bareilly-243122, Uttar Pradesh
3. National Dairy Research Institute, Karnal-132001, Haryana
4. Central Institute of Fisheries Education, Mumbai-400061, Maharashtra
5. Allahabad Agricultural Institute, Allahabad-211007, Uttar Pradesh

Central Universities with Agriculture Faculty

1. Banaras Hindu University, Varanasi, U.P.
2. Aligarh Muslim University, Aligarh, U.P.
3. Vishwa Bharti, Shantiniketan, West Bengal
4. Nagaland University, Medizipherma, Nagaland

PRACTICE SET

Q. 1. The recommended safe limits of land slope in border for medium loam soils are:

(A) 0.20 to 0.40%. (B) 0.05 to 0.10%
(C) 0.10 to 0.2% (D) 0.10 to 0.5%

Q. 2. The discounting procedure which is satisfactory for ranking independent projects

(A) Benefit-cost ratio
(B) Net benefit investment ratio
(C) Internal rate of return
(D) Net present worth

Q. 3. The most efficient equipment for making ridges in the layout of fields for irrigation borders large size check basins is:

(A) Bund former
(B) A-frame ridger
(C) Disc ridger
(D) Double mould board plough

Q. 4. The current price is:

(A) The value or price as recorded
(B) The value or price that is expected to occur under normal conditions
(C) Shadow price
(D) Imputed price

Q. 5. The most commonly adopted land levelling design in small scale jobs is:

(A) Contour adjustment method
(B) Profile method
(C) Plane method
(D) Plan inspection method

Q. 6. The most appropriate wooden equipment for land smoothening operation is:

(A) Leveller board
(B) Buck scraper
(C) Float
(D) U-leveller

Q. 7. The most appropriate mechanically operated equipment for medium scale land grading and smoothening is:

(A) Terracer blade
(B) Land plane
(C) Two-wheeled automatic leveller
(D) Carrier-type scraper

Q. 8. In a land plane the scraper blade is provided at:

(A) The front of the frame
(B) About the centre of the frame
(C) The back of the frame
(D) One-third distance from the front of the frame

Q. 9. The most significant component in the water charges levied by the government is:

(A) Betterment levy
(B) Incremental land revenue
(C) Water rate
(D) Irrigation cess

Q. 10. The side slope of the dyke in a contour bench levelling job 1s kept about:

(A) 1:1 (B) 2:1
(C) 1.5:1 (D) 2:5:1

Q. 11. In case of irrigation projects, the usually adopted value of the length of the project period economic analysis is:

(A) 15 years (B) 25 years
(C) 40 years (D) 50years

Q. 12. The discounting procedure suitable for ranking mutually exclusive projects is:

(A) Benefit-cost ratio
(B) Net benefit investment ratio
(C) Internal rate of return
(D) Net present worth

Q. 13. Tensiometer installed in soils measure:

(A) Osmotic suction of soil moisture
(B) Capillary potential of the soil

(C) Soil moisture content
(D) Soil moisture constants

Q. 14. Though not an essential micro-nutrient of plants, the availability of selenium in the harvested produce of the crop is highly desirable in:
(A) Cereal crops (B) Pulses
(C) Oilseeds (D) Fodder crops

Q. 15. The diameter of the intake pipe of a lift scheme flowing partially full is computed using:
(A) Darcy Weisbach formula
(B) William Razen formula
(C) Manning's formula
(D) Chezy's formula

Q. 16. Transfer of heat to gram may take place directly from surface by means of:
(A) Latest heat
(B) Sensible heat
(C) Heat conductivity
(D) Heat content

Q. 17. Field performance index of a machine is also known as its:
(A) Actual field capacity
(B) Theoretical field capacity
(C) Field efficiency
(D) All above

Q. 18. First prototype of zero seeddrill/No-till drill was developed at:
(A) Punjab Agricultural University Ludhiana
(B) Govind vallabh Pant University of Agriculture and Technology, Pant Nagar
(C) Choudhary Charan Singh University of Agriculture, Hissar
(D) Indian Agricultural Research Institute, New Delhi

Q. 19. The density of biogas slurry is about:
(A) 1080 kg/m^3
(B) 500 kg/m^3
(C) 1080 gm/m^3
(D) 500 gm/m^3

Q. 20. The size of Janta type biogas plant is limited to:
(A) 8 Cubic metre
(B) 12 Cubic metre
(C) 10 Cubic metre
(D) 15 Cubic metre

Q. 21. In case of semi-conductor, following are responsible for current movement
(A) Free electron
(B) Free electrons and holes
(C) Only holes
(D) None of these

Q. 22. Gravity separator works on the principle of:
(A) Gravity
(B) Drag
(C) Both (A) and (B)
(D) None of these

Q. 23. In the meandering course of a river in alluvial formation, the intake of a lift scheme is to be located on:
(A) Concave side of the river
(B) Convex side of the river
(C) At a point where the water flow is deep
(D) On a straight length of the river

Q. 24. Itake structures for lift irrigation from rivers in delta regions generally comprise of:
(A) Portable pumping sets
(B) Floating barges and their adaptations
(C) Specially designed intake structures suiting the river flow stage
(D) A combination of (B) and (C)

Q. 25. The most important requirement in the installation of RCC intake pipe is:
(A) Providing the required bed slope in the pipeline
(B) Ensuring watertight pipe joints
(C) Provision of man-holes where the length of the pipeline exceeds about 100 m
(D) Provision of suitable concrete cradles as bottom support for the pipes

Q. 26. The survey used for various agricultural engineering interventions on the farm is:

(A) Cadastral survey
(B) Topographic survey
(C) Revenue survey.
(D) Agricultural survey

Q. 27. Grouping of wells should not be made at the interval less than:

(A) 100 m (B) 300 m
(C) 400 m (D) 200 m

Q. 28. A streamlined body is defined as a body about which:

(A) The flow is laminar.
(B) The flow is along the streamlines
(C) The flow separation is suppressed.
(D) The drag is zero

Q. 29. The Pitot-static tube measures:

(A) Static pressure.
(B) Difference in static and dynamic pressures
(C) Dynamic pressure.
(D) Difference in total and static pressures

Q. 30. The benefits of implementing integrated watershed development programme are:

(A) Soil conservation.
(B) Rainwater harvesting and conservation
(C) Regeneration of natural vegetation.
(D) (A), (B) and (C)

Q. 31. Integrated Watershed Management Programme' is implemented by:

(A) Ministry of Agriculture
(B) Ministry of Environment and Forest
(C) Ministry of Rural Development.
(D) Ministry of Water Resources

Q. 32. Irrigation is useful in a situation where

(A) Saline water is used for irrigation
(B) Soil is heavy clay to permit high capillary rise
(C) Capillary movement in the root zone is rapid
(D) No hard pan is present below the root zone

□